세상에서 가장 쉬운 과학 수업

양자광학

세상에서 가장 쉬운 과학 수업

ⓒ 정완상, 2025

초판 1쇄 인쇄 2025년 7월 7일
초판 1쇄 발행 2025년 7월 17일

지은이 정완상
펴낸이 이성림
펴낸곳 성림북스

책임편집 최윤정
디자인 쏘울기획

출판등록 2014년 9월 3일 제25100-2014-000054호
주소 서울시 은평구 연서로3길 12-8, 502
대표전화 02-356-5762
팩스 02-356-5769
이메일 sunglimonebooks@naver.com

ISBN 979-11-93357-70-5 03400

* 책값은 뒤표지에 있습니다.
* 이 책의 판권은 지은이와 성림북스에 있습니다.
* 이 책의 내용 전부 또는 일부를 재사용하려면 반드시 양측의 서면 동의를 받아야 합니다.

노벨상 수상자들의 **오리지널 논문**으로 배우는 과학

세상에서 가장 쉬운 과학 수업

양자광학

정완상 지음

빛의 입자설과 파동설부터 양자컴퓨터까지
미래 기술을 이끌어갈 빛의 신비로운 세계를 탐구하다

성림원북스

CONTENTS

추천사   008
천재 과학자들의 오리지널 논문을 이해하게 되길 바라며   014
양자광학의 창시자 글라우버_애슈킨 박사 깜짝 인터뷰   017

첫 번째 만남
고전광학 / 021

입자설과 파동설_날아가는 야구공 vs 출렁거리는 파도   022
파동설의 승리_간섭무늬와 회절 현상   025
전자기파의 발견_눈에 보이지 않는 빛을 찾다   031

두 번째 만남
레이저의 탄생 / 041

광자의 탄생_빛을 이루는 입자   042
아인슈타인의 유도방출 이론_레이저의 기본 원리   043
양자역학의 탄생_전자라는 양자가 만족하는 물리학   046
메이저의 발명_강렬하고 정밀한 빔을 만들다   052
레이저의 발명_증폭된 가시광선의 빔   061
레이저 응용_천문학에서 의학까지   066

세 번째 만남
광통신의 역사 / 069

그레이엄 벨_광통신을 발명하다   070
포토폰의 발명_거울에 반사되는 빛으로   077
광섬유의 발견_유리 막대에 갇힌 빛   083
저손실 광섬유의 발명과 광통신_더 효율적으로 많은 정보를   087

네 번째 만남
양자광학의 탄생 / 093

광자의 생성·소멸 연산자_광자 한 개를 어디로 보낼까?   094
벡터 대수학_전기장과 자기장을 연산자로 바꾸기 위해   102

맥스웰 방정식_전기장과 자기장을 묘사하는 복잡한 방정식   107
파동방정식과 전자기파_파동은 어떻게 나타낼까?   110
전자기파_벡터퍼텐셜이 만족하는 파동방정식에서   117
전자기장의 양자화_입체 속의 전자기파   121

다섯 번째 만남
원자와 빛의 상호작용 / 127

제인스-커밍스 모형_바닥상태와 첫 번째 들뜬상태   128
라비 진동_원자 속 전자가 두 에너지 준위 사이에서 진동한다   135

여섯 번째 만남
결맞는 상태 / 143

글라우버_로스앨러모스의 최연소 과학자   144
결맞는 상태_빛의 선명도가 최대가 되다   146
결맞는 상태의 성질_광자수의 확률, 기댓값, 표준편차   152
반응집광_방출되는 광자가 분리되는 현상   159

일곱 번째 만남

양자광 기술 / 161

애슈킨의 광학 핀셋_레이저 빔으로 미세한 물체를 붙잡다 — 162

주파수 빗의 발명_일정한 주파수 간격의 날카로운 스펙트럼선 — 166

레이저 냉각_원자의 속력을 느리게 만들다 — 169

메타물질_투명 망토의 비밀 — 174

반응집광과 양자컴퓨터_양자광학의 활용과 미래 — 181

만남에 덧붙여 / 185

Comparison of Quantum and Semiclassical Radiation Theories with Application to the Beam Maser_제인스-커밍스 논문 영문본 — 186

Coherent and Incoherent States of the Radiation Field_글라우버 논문 영문본 — 207

Controlling Electromagnetic Fields_펜드리 논문 영문본 — 230

위대한 논문과의 만남을 마무리하며 — 233

이 책을 위해 참고한 논문들 — 236

수식에 사용하는 그리스 문자 — 240

노벨 물리학상 수상자들을 소개합니다 — 241

과학을 처음 공부할 때 이런 책이 있었다면 얼마나 좋았을까

남순건(경희대학교 이과대학 물리학과 교수 및 전 부총장)

21세기를 20여 년 지낸 이 시점에서 세상은 또 엄청난 변화를 맞이하리라는 생각이 듭니다. 100년 전 찾아왔던 양자역학은 반도체, 레이저 등을 위시하여 나노의 세계를 인간이 이해하도록 하였고, 120년 전 아인슈타인에 의해 밝혀진 시간과 공간의 원리인 상대성이론은 이 광대한 우주가 어떤 모습으로 만들어져 왔고 앞으로 어떻게 진화할 것인가를 알게 해주었습니다. 게다가 우리가 사용하는 모든 에너지의 근원인 태양에너지를 핵융합을 통해 지구상에서 구현하려는 노력도 상대론에서 나오는 그 유명한 질량-에너지 공식이 있기에 조만간 성과가 있을 것이라 기대하게 되었습니다.

앞으로 올 22세기에는 어떤 세상이 펼쳐질지 매우 궁금합니다. 특히 인공지능의 한계가 과연 무엇일지, 또한 생로병사와 관련된 생명의 신비가 밝혀져 인간 사회를 어떻게 바꿀지, 우주에서는 어떤 신비로움이 기다리고 있는지, 우리는 불확실성이 가득한 미래를 향해 달려가고 있습니다. 이러한 불확실한 미래를 들여다보는 유리구슬 역할을 하는 것이 바로 과학적 원리들입니다.

지난 백여 년간 과학에서의 엄청난 발전들은 세상의 원리를 꿰뚫어보았던 과학자들의 통찰을 통해 우리에게 알려졌습니다. 이런 과학 발전을 가능하게 한 영웅들의 생생한 숨결을 직접 느끼려면 그들이 썼던 논문들을 경험해보는 것이 좋습니다. 그런데 어느 순간 일반인과 과학을 배우는 학생들은 물론, 그 분야에서 연구를 하는 과학자들마저 이런 숨결을 직접 경험하지 못하고 이를 소화해서 정리해놓은 교과서나 서적들을 통해서만 접하고 있습니다. 창의적인 생각의 흐름을 직접 접하는 것은 그런 생각을 했던 과학자들의 어깨 위에서 더 멀리 바라보고 새로운 발견을 하고자 하는 사람들에게 매우 중요합니다.

저자인 정완상 교수가 새로운 시도로써 이러한 숨결을 우리에게 전해주려 한다고 하여 그의 30년 지기인 저는 매우 기뻤습니다. 그는 대학원생 때부터 당시 혁명기를 지나면서 폭발적인 발전을 하고 있던 끈 이론을 위시한 이론물리학 분야에서 가장 많은 논문을 썼던 사람입니다. 그리고 그러한 에너지가 일반인들과 과학도들을 위한 그의 수많은 서적을 통해 이미 잘 알려져 있습니다. 저자는 이번에 아주 새로운 시도를 하고 있고 이는 어쩌면 우리에게 꼭 필요했던 것일 수 있습니다. 대화체로 과학의 역사와 배경을 매우 재미있게 설명하고, 그 배경 뒤에 나왔던 과학 영웅들의 오리지널 논문들을 풀어간 것입니다. 과학사를 들려주는 책들은 많이 있으나 이처럼 일반인과 과학도의 입장에서 질문하고 이해하는 생각의 흐름을 따라 설명한 책은 없습니다. 게다가 이런 준비를 마친 후에 아인슈타인 같은 영웅들의 논문을 원래의 방식과 표기를 통해 설명하는 부분은 오랫동안 과학을 연구해온 과학자에게도 도움을 줍니다.

이 책을 읽는 독자들은 복 받은 분들일 것이 분명합니다. 제가 과학을 처음 공부할 때 이런 책이 있었다면 얼마나 좋았을까 하는 생각이 듭니다. 정완상 교수는 이제 새로운 형태의 시리즈를 시작하고 있습니다. 독보적인 필력과 독자에게 다가가는 그의 친밀성이 이 시리즈를 통해 재미있고 유익한 과학으로 전해지길 바랍니다. 그리하여 과학을 멀리하는 21세기의 한국인들에게 과학에 대한 붐이 일기를 기대합니다. 22세기를 준비해야 하는 우리에게는 이런 붐이 꼭 있어야 하기 때문입니다.

미래를 밝힐 양자광학, 가장 쉬운 길라잡이를 만나다

김승진(세종과학예술영재학교 물리 교사)

21세기 과학기술의 눈부신 발전 속에서 '양자'라는 단어는 더 이상 전문가만의 것이 아닙니다. 미래 사회의 패러다임을 바꿀 양자컴퓨터, 양자통신 등의 이야기를 들을 때마다 우리 학생들의 눈은 호기심으로 빛납니다. 동시에 양자물리학이라는 거대한 산 앞에서 지레 겁을 먹기도 합니다. 바로 이런 학생들에게 용기와 영감을 불어넣을 귀한 책, 정완상 교수님의 《세상에서 가장 쉬운 과학 수업 양자광학》을 소개하게 되어 매우 기쁩니다.

이 책은 단순히 양자광학의 개념을 나열하는 데 그치지 않습니다. 저자는 독자들을 시간 여행으로 초대하여, 빛이 입자인지 파동인지 치열하게 논쟁했던 고전광학의 시대로 먼저 데려갑니다. 이어서 아인슈타인의 유도방출 이론이 레이저라는 혁신적인 빛을 탄생시킨 과정, 유리 막대에 빛을 가두어 정보의 바다를 열어젖힌 광섬유의 발견에 이르기까지, 한 편의 흥미진진한 과학 드라마를 펼쳐 보입니다. 이러한 역사적 배경 위에서 독자들은 자연스럽게 양자광학의 핵심 질문들과 마주하게 됩니다.

무엇보다 이 책의 백미는 저자가 서문에서 밝힌 '이제는 독자들의 수준도 많이 높아졌으니 수식을 피하지 말고 천재 과학자들의 오리지널 논문을 이해하길 바랐습니다'라는 열망이 담긴 부분입니다. 양자광학의 핵심이라 할 수 있는 광자의 생성·소멸 연산자, 빛과 원자의 상호작용을 다룬 제인스–커밍스 모형, 빛의 최대 간섭을 설명하는 글라우버의 결맞는 상태 등 복잡한 개념들을 '고등학교 정도의 수식을 이해하는 청소년과 일반 독자'의 눈높이에 맞게 풀어내는 저자의 노력은 실로 감탄을 자아냅니다. 이는 학생들이 단순히 지식을 습득하는 것을 넘어, 위대한 과학자들이 문제를 어떻게 정의하고 해결해 나갔는지 그 창의적인 생각의 흐름을 간접적으로나마 접하는 소중한 기회가 될 것입니다.

또한 이 책은 '전자기장의 양자화' 같은 근본적인 개념 설명부터 시작하여, 이러한 기초가 어떻게 현대 과학기술의 눈부신 성과로 이어지는지를 생생하게 보여줍니다. 레이저 빔으로 미세 입자를 다루는 '애슈킨의 광학 핀셋', 원자시계의 정밀도를 극한으로 끌어올린 '주파수 빗의 발명', 빛의 경로를 조종하는 '메타물질', 그리고 미래 컴퓨팅의 핵심인 '양자컴퓨터'에 이르기까지 양자광학의 다양한 응용 사례들을 소개합니다. 이처럼 주요 연구 내용과 관련 연구자들을 폭넓게 조명하며, 특히 '노벨 물리학상 수상자들을 소개합니다' 코너를 통해 학생들의 지적 탐구심을 자극하고 과학자로서의 꿈을 키울 수 있도록 안내합니다.

《세상에서 가장 쉬운 과학 수업 양자광학》은 양자 시대를 살아갈 우리 학생들에게 과학적 소양을 넓히고 미래를 준비하는 데 더없이 좋은 길잡이가 될 것입니다. 저자의 바람처럼 이 책을 통해 많은 학생이 과학에 대한 두려움을 떨치고, '이 책의 30% 이상 이해한다면 그 사람은 대단하다고 봅니다'라는 격려에 힘입어 즐겁게 양자광학의 세계를 탐험하기를 기대합니다.

천재 과학자들의 오리지널 논문을
이해하게 되길 바라며

　　사람들은 과학 특히 물리학 하면 너무 어렵다고 생각하지요. 제가 외국인들을 만나서 얘기할 때마다 신선하게 느끼는 점이 있습니다. 그들은 고등학교까지 과학을 너무 재미있게 배웠다고 하더군요. 그래서인지 과학에 대해 상당한 지식을 가진 사람들이 많았습니다. 그 덕분에 노벨 과학상도 많이 나오는 게 아닐까 생각해요. 우리나라는 노벨 과학상 수상자가 한 명도 없습니다. 이제 청소년과 일반 독자의 과학 수준을 높여 노벨 과학상 수상자가 매년 나오는 나라가 되게 하고 싶다는 게 제 소망입니다.

　　그동안 양자역학과 상대성이론에 관한 책은 전 세계적으로 헤아릴 수 없을 정도로 많이 나왔고 앞으로도 계속 나오겠지요. 대부분의 책은 수식을 피하고 관련된 역사 이야기들 중심으로 쓰여 있어요. 제가 보기에는 독자를 고려하여 수식을 너무 배제하는 것 같았습니다. 이제는 독자들의 수준도 많이 높아졌으니 수식을 피하지 말고 천재 과학자들의 오리지널 논문을 이해하길 바랐습니다. 그래서 앞으로 도래할 양자(量子, quantum)와 상대성 우주의 시대를 멋지게 맞이하도록 도우리라는 생각에서 이 기획을 하게 된 것입니다.

원고를 쓰기 위해 논문을 읽고 또 읽으면서 어떻게 이 어려운 논문을 독자들에게 알기 쉽게 설명할까 고민했습니다. 여기서 제가 설정한 독자는 고등학교 정도의 수식을 이해하는 청소년과 일반 독자입니다. 물론 이 시리즈의 논문에 그 수준을 넘어서는 내용도 나오지만 고등학교 수학만 알면 이해할 수 있도록 설명했습니다. 이 책을 읽으며 천재 과학자들의 오리지널 논문을 얼마나 이해할지는 독자들에 따라 다를 거라 생각합니다. 책을 다 읽고 100% 혹은 70%를 이해하거나 30% 미만으로 이해하는 독자도 있을 것입니다. 제 생각으로는 이 책의 30% 이상 이해한다면 그 사람은 대단하다고 봅니다.

여기에서는 양자광학에 대한 세 개의 논문(제인스-커밍스 모형 논문, 글라우버의 결맞는 상태 논문, 펜드리의 투명 망토 논문)을 다루었습니다. 우선 제인스-커밍스 모형 논문이 나오기까지 과학자들의 업적을 소개했습니다. 빛의 입자설과 파동설의 논쟁, 광통신의 역사, 메이저와 레이저의 발명에 얽힌 이야기들이 그것입니다. 역사적으로 재미있는 일화를 곁들여 독자들이 읽는 데 지루함이 없게 했습니다. 그리고 광자의 생성과 소멸 연산자를 다룬 디랙의 업적과 전자기파를 양자화하는 과정에 대해 이야기했습니다. 또한 제인스-커밍스 논문에서 묘사한 빛과 원자의 상호작용을 다루었습니다. 이 논문을 통해 원자가 빛을 흡수하거나 방출할 때 원자 속 전자의 상태가 어떻게 변화되는지를 이해할 수 있습니다.

다음으로 양자광학의 최소 불확정성 상태인 글라우버의 결맞는 상태를 설명했습니다. 양자광의 특징 중 하나인 반응집성도 다루었습니다. 마지막으로 현대 사회에서 양자광학의 활용을 살펴보았습니다. 이러한 배경 설명을 통해 여러분은 양자광학이 무엇인지 이해할 수 있을 것입니다.

   〈노벨상 수상자들의 오리지널 논문으로 배우는 과학〉 시리즈는 많은 이에게 도움을 줄 수 있다고 생각합니다. 과학자가 꿈인 학생과 그의 부모, 어릴 때부터 수학과 과학을 사랑했던 어른, 양자역학과 상대성이론을 좀 더 알고 싶은 사람, 아이들에게 위대한 논문을 소개하려는 과학 선생님, 반도체나 양자암호 시스템, 우주 항공 계통 등의 일에 종사하는 직장인, 〈인터스텔라〉를 능가하는 SF 영화를 만들고 싶어 하는 영화 제작자나 웹툰 작가 등 많은 사람들에게 이 시리즈를 추천합니다.

진주에서 정완상 교수

양자광학의 창시자 글라우버 _ 애슈킨 박사 깜짝 인터뷰

글라우버의 반응집광 연구

기자 오늘은 글라우버의 결맞는 상태와 반응집광 연구에 관해 애슈킨 박사와 인터뷰를 진행하겠습니다. 애슈킨 박사는 2018년 광학 핀셋으로 노벨 물리학상을 수상한 분이지요. 애슈킨 박사님, 나와 주셔서 감사합니다.

애슈킨 제가 제일 존경하는 과학자인 글라우버의 논문에 관한 내용이라 만사를 제치고 달려왔습니다.

기자 글라우버를 양자광학의 창시자로 보는 이유는 무엇일까요?

애슈킨 양자광학은 빛의 양자적 특성을 다루는 학문입니다. 이렇게 양자적 특성을 가진 빛을 양자광이라고 하지요. 양자광의 대표적인 특징 중 하나가 반응집성입니다. 글라우버는 이 반응집성을 최초로 연구했지요.

기자 반응집성이 뭐죠?

애슈킨 광자의 반응집성(Photon antibunching)은 광원에서 방출될 때 개별 광자가 일시적으로 분리되는 양자역학적 현상을 말합니다. 이렇게 방출되는 빛을 반응집광이라 하고, 반응집광과 반대로 광자들이 몰려다니는 행동을 보이는 빛을 응집광이라고 부릅니다.

기자 흥미롭군요.

글라우버 이전의 양자광학 연구

기자 글라우버 이전에 양자광학을 연구한 과학자들의 이야기를 들려주세요.

애슈킨 빛을 양자로 처음 다룬 건 1900년 막스 플랑크입니다. 양자의 특성을 가진 빛을 이루는 알갱이 하나하나를 광자라고 부르지요. 광자의 존재는 아인슈타인의 광전효과에 의해 밝혀졌고, 그 후 디랙이 광자에 대한 양자역학을 만들었습니다. 광자의 개념을 이용해 탄생한 위대한 발명품이 바로 메이저와 레이저입니다.

기자 빛과 물질의 관계는 누가 알아냈나요?

애슈킨 그것은 제인스와 커밍스가 찾아냈습니다. 제인스-커밍스 모형이라고 불리지요.

기자 그렇군요.

글라우버의 1963년 논문 개요

기자 글라우버의 1963년 논문에는 무슨 내용이 들어 있나요?

애슈킨 결맞는 상태와 반응집광에 대한 연구를 다루었습니다.

기자 결맞는 상태는 무엇인가요?

애슈킨 광자를 양자역학적으로 다룰 때 광자가 최소 불확정성원리를 만족하도록 하는 상태이지요. 글라우버는 다양한 광자수 상태의 중첩으로 어떤 상태를 만든 후 이에 대한 최소 불확정성원리를 요구해 결맞는 상태를 만드는 데 성공했습니다. 그는 결맞는 상태 속에서 특정 광자수를 가진 상태를 발견할 확률이 푸아송 분포를 따르는 것도 발견했습니다.

기자 더 자세히 알아보고 싶네요.

글라우버의 1963년 논문이 일으킨 파장

기자 글라우버의 1963년 논문은 무슨 변화를 가져왔나요?

애슈킨 양자광 기술을 만드는 혁명을 일으켰습니다.

기자 어떤 기술을 말하죠?

애슈킨 레이저를 이용한 냉각 기술로 원자를 냉각시켜 원자시계를 만들 수 있는 기술이지요. 또한 양자광학은 양자정보과학에서 이용됩니다. 광자를 가지고 큐비트를 만들 때 반응집광 개념을 사용하지요. 빛의 반응집성은 전송 오류와 보안 위반을 줄여 양자암호화 프로토콜의 보안을 강화합니다. 그래서 전송된 정보를 가로채거나 디코딩하는 것을 어렵게 만듭니다.

기자 정말 많은 곳에 쓰이는군요. 지금까지 글라우버의 결맞는 상태와 반응집광 연구에 대해 애슈킨 박사의 이야기를 들어 보았습니다.

첫 번째 만남

•

고전광학

입자설과 파동설 _ 날아가는 야구공 vs 출렁거리는 파도

정교수 이제 우리는 양자광학에 대한 이야기를 시작하려고 해.

물리군 양자와 광학이 합쳐진 이론인가요?

정교수 맞아. 광학은 빛을 다루는 물리학인데 고전물리로 빛을 다루는 광학을 고전광학, 양자론으로 빛을 다루는 광학을 양자광학이라고 생각하면 돼.

물리군 고전광학에서는 빛을 어떻게 다루죠?

정교수 과학자들은 오랫동안 논쟁을 벌였어. 빛이 날아가는 야구공 같은 입자인지, 아니면 출렁거리는 파도 같은 파동인지 말이야.

빛의 입자설을 처음 주장한 사람은 영국의 뉴턴이다. 그는 1672년 영국 왕립학회에 보낸 편지에서 빛은 동일한 입자들로 이루어진 광선이라고 주장했다. 이것을 뉴턴의 빛의 입자설이라고 부른다.

1666년 뉴턴은 프리즘을 통한 빛의 분산을 알아냈는데, 이를 바탕으로 입자설을 주장했다. 그는 백색광이 서로 다른 색상의 입자들로 이루어져 있다고 보았다. 그리고 이런 내용을 토대로 빛의 반사와 굴절을 설명했다. 빛의 입자설에 대한 뉴턴의 아이디어는 1704년에 그가 쓴 《광학(Opticks)》에 집대성되었다.

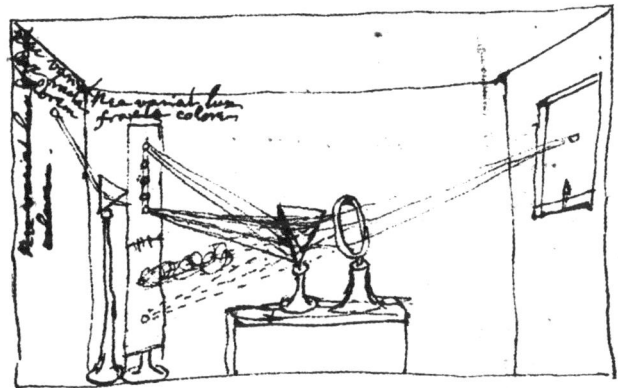

프리즘 실험에 대한 뉴턴의 스케치(출처: Patricia Fara/the Royal Society)

비슷한 시기에 네덜란드의 하위헌스는 빛의 파동설을 주장했다. 그는 1678년에 쓴 《빛에 관한 논고(Traité de la Lumière)》라는 책에서 그 유명한 하위헌스의 원리를 발표했다.

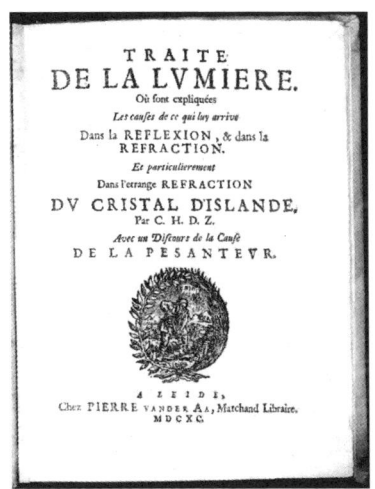

《빛에 관한 논고》

물리군   하위헌스의 원리가 뭔가요?

정교수   파동이 전파하는 모습을 결정하는 원리야. 바다에서 밀려오는 파도를 관찰하면 파의 높은 부분, 즉 마루를 이루는 곡선이 그 형태를 조금씩 바꾸면서 움직이는 것을 알 수 있지?

파동의 마루를 이은 곡선을 파면이라고 불러. 어느 순간의 파면이 주어지면 다음 순간의 파면은 주어진 파면상의 각 점에서 발생하는 구면파들에 공통으로 접하는 면이 된다는 것이 하위헌스의 원리야.

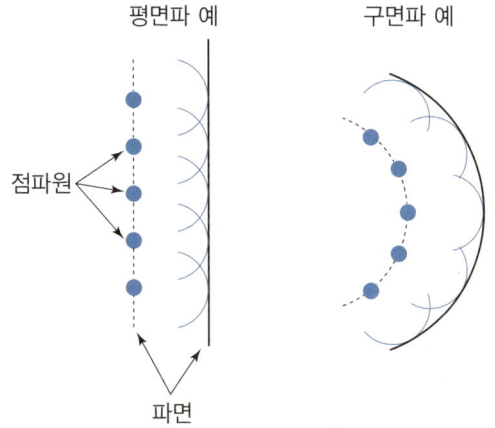

물리군   입자설과 파동설은 완전히 다른 이론이군요.

정교수   그래. 하지만 뉴턴과 하위헌스의 시대에는 사람들이 뉴턴의 입자설을 더 지지했지. 아마도 그의 명성이 훨씬 더 높았기 때문인 듯해.

물리군   그렇겠네요.

파동설의 승리 _ 간섭무늬와 회절 현상

정교수   19세기로 들어오면서 빛의 입자설보다는 파동설에 유리한 증거들이 등장했어. 그중에 제일 대표적인 것은 영의 실험이야. 영국의 의사이자 물리학자인 영(Thomas Young, 1773~1829)은 그 유명한 빛의 이중 슬릿 실험을 했지.

1803년 영은 두 개의 슬릿(작은 구멍)을 통과한 빛들이 간섭을 일으켜 스크린에 밝고 어두운 무늬를 만드는 것을 알아냈다. 이 실험은 하위헌스의 원리로 잘 설명되었다. 반면 이러한 간섭무늬를 뉴턴의 입자설로는 절대로 설명할 수 없었다.

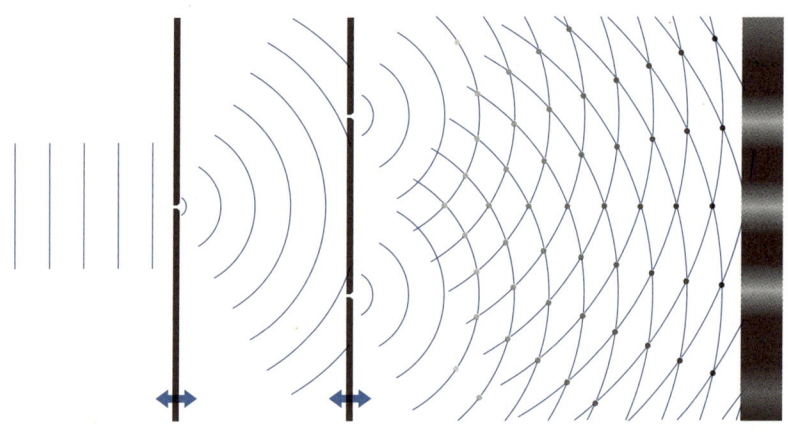

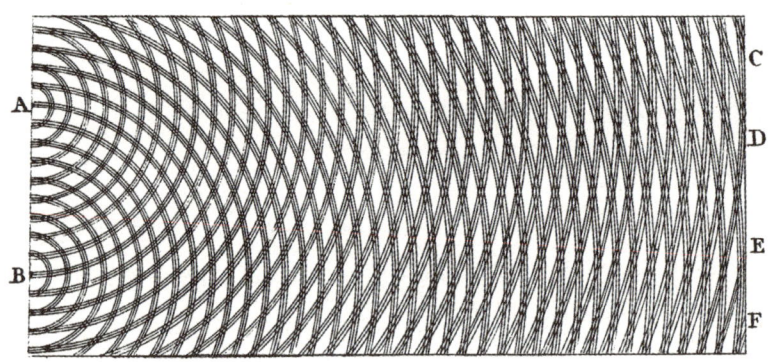

빛의 간섭에 대한 영의 자필 스케치

물리군 빛의 성질 중에서 입자설로 설명할 수 없는 게 또 있나요?

정교수 물론이야. 회절 현상도 빛의 입자설로는 설명할 수 없어. 이 문제는 프랑스의 물리학자 프레넬이 주로 연구했지.

프레넬(Augustin-Jean Fresnel, 1788~1827)

프레넬은 1788년 프랑스 노르망디 지방의 브로글리에서 태어났다. 그는 집에서 어머니로부터 교육을 받았다. 몸이 약한 프레넬은 다른 아이들에 비해 암기력이 떨어지는 편이었다. 하지만 아홉 살 때 나뭇가지를 가지고 장난감 활과 총을 만들기도 했다.

브로글리(출처: Codepem/Wikimedia Commons)

1801년 프레넬은 캉의 에콜 상트랄에 입학해 처음으로 과학을 접했다. 1804년에는 에콜 폴리테크니크에 들어가 과학과 공학을 공부했다. 이 시기에 그는 친구를 거의 사귀지 못했고, 건강도 계속 나빴다. 그럼에도 불구하고 그림과 기하학에 뛰어난 재능을 보였다.

1806년에 에콜 폴리테크니크를 졸업한 프레넬은 국립 교량 도로 학교에 입학했다. 거기서 토목공학을 전공하고 1809년에 졸업해 프랑스 기술 군단의 엔지니어로 일했다.

프레넬은 1814년부터 광학에 관심을 두었다. 당시 그는 하위헌스의 파동설을 지지했고 영의 실험 내용을 흥미 있게 읽었다. 그가 주목한 연구 분야는 빛의 회절이었다.

회절(diffraction)은 대표적인 파동 현상 중 하나이다. 파동이 장애물이나 좁은 틈을 통과할 때, 그 뒤편(그림자 부분)까지 전파하는 현상을 회절이라고 부른다. 입자는 틈을 통과하면 그 틈을 지나 직선으로 진행한다. 이와 달리 파동의 경우, 틈을 지나는 직선 경로뿐 아니라 그 주변의 일정 범위까지 돌아 들어간다. 이처럼 파동이 입자로서는 도저히 갈 수 없는 영역에 휘어져 도달하는 현상이 회절이다.

프레넬은 입자설로는 빛의 회절을 절대로 설명할 수 없는 걸 알고, 회절 이론을 연구하기로 결심했다. 그는 회절의 정도가 틈의 크기와 파장에 영향을 받음을 알아냈다. 특히 틈의 크기에 비해 파장이 길수록 회절이 더 많이 일어나는 것을 밝혔다. 즉, 파장이 일정할 때 틈의 크기가 작을수록 회절이 잘 일어나, 파면이 직선이었던 파동이 좁은 틈을 지나면 반원에 가까운 모양으로 퍼진다는 것이다.

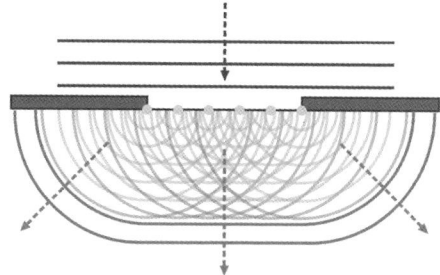

회절(출처: Arne Nordmann (norro))

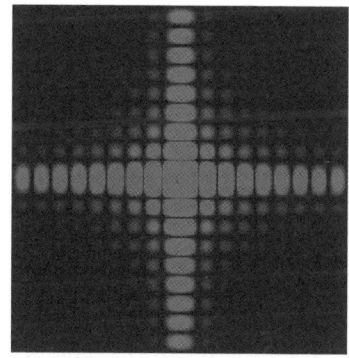

사각형 구멍에서의 회절

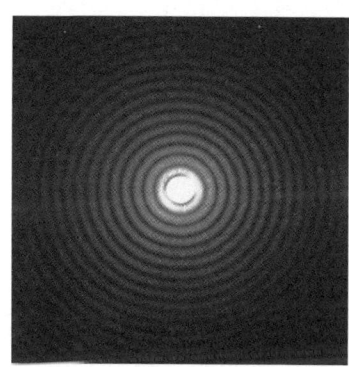

원형 구멍에서의 회절
(출처: Wisky/Wikimedia Commons)

프레넬은 구멍에 의해 회절된 빛의 밝기를 모든 지점에서 구할 수 있는 공식을 만들었다. 이 공식은 실험과 잘 맞았다.

드디어 프레넬의 업적을 과학자들에게 알릴 기회가 찾아왔다. 1819년 파리 과학 아카데미는 빛의 회절을 가장 잘 설명하는 과학자에게 상을 준다고 현상공모를 했다. 프레넬은 자신의 연구 결과를 파리 과학 아카데미에 제출했다. 심사 위원 5명 중 3명이 빛의 입자설을 지지했음에도 그는 당당히 1등을 차지했다.

또한 프레넬은 등대 광원에서 나오는 빛을 더 잘 모으는 렌즈를 발명했는데, 이를 프레넬 렌즈라고 부른다. 프레넬 렌즈는 다른 렌즈들과 달리 평평한 모양에, 가장자리는 마치 톱니바퀴처럼 깎여 있었다.

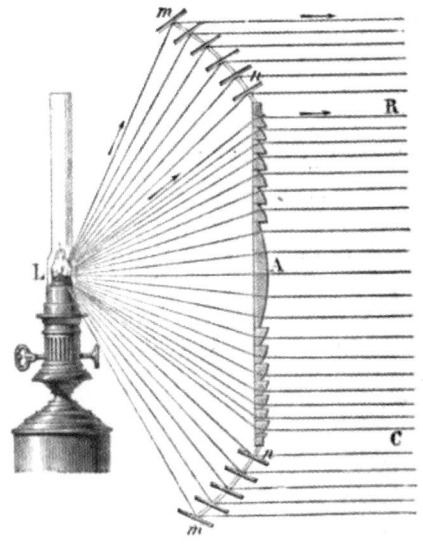

프레넬 렌즈

물리군 빛의 회절로 파동설이 완전한 승리를 거두었군요.

정교수 그런 셈이야.

전자기파의 발견 _ 눈에 보이지 않는 빛을 찾다

정교수 1867년 스코틀랜드의 물리학자 맥스웰(James Clerk Maxwell)이 전기와 자기에 관한 통일 방정식인 맥스웰 방정식(Maxwell's equations)을 발표했다네. 이로써 빛의 파동설이 압승을 거두지. 맥스웰은 이 방정식을 통해 전자기파의 속도가 빛의 속도라는 걸 알아냈어. 그러니까 빛이 전자기파라는 파동임을 증명한 거야. 그리고 1887년 독일의 물리학자 헤르츠(Heinrich Hertz)는 자신의 실험실에서 전자기파를 생성하는 데 성공했지.

전자기파는 눈에 보이는 것도 있고, 보이지 않는 것도 있는데 이는 전자기파의 파장에 의해 결정된다. 눈에 보이는 전자기파를 가시광선이라고 부른다. 각각의 파장에 대한 전자기파의 이름은 다음과 같다.

파장	이름
< 0.01nm	감마선
0.01~10nm	엑스선
10~400nm	자외선
400~750nm	가시광선

파장	이름
750nm~1mm	적외선
1mm~1m	마이크로파
≥1m	라디오파

여기서

$1m = 10^6 \mu m$(마이크로미터)

$1m = 10^9 nm$(나노미터)

이다.

물리군 눈에 안 보이는 전자기파도 맥스웰과 헤르츠가 발견했나요?

정교수 그렇지는 않아. 맥스웰과 헤르츠가 발견한 전자기파는 마이크로파와 라디오파 정도야. 엑스선과 감마선은 20세기 초에 방사선을 연구하던 과학자들이 찾아냈지. 적외선과 자외선은 맥스웰이 등장하기 훨씬 전에 발견되었어. 그럼 적외선 발견의 역사부터 알아볼게.

적외선은 천왕성을 발견한 허셜이 처음 찾아냈다. 1800년 초, 허셜은 햇빛을 통과시키기 위해 여러 필터를 테스트하고 있었다. 이때 다양한 색상의 필터에서 열을 생성하는 양이 제각각 달라지는 걸 알아냈다. 그는 온도계를 사용하여 빛의 다양한 색상에 따른 열을 측정하고자 했다. 프리즘에 빛을 통과시키고, 그 과정에서 가시광선 스펙트럼의 빨간색 끝 바로 너머를 측정했다. 이 실험을 통해 적색광보다 1도 높은 온도를 감지했다. 추가 실험으로 허셜은 적색광 너머에 보이지 않는 형태의 빛이 있어야 한다는 결론을 내렸고, 1800년 4월에 이 결과를 발표했다.

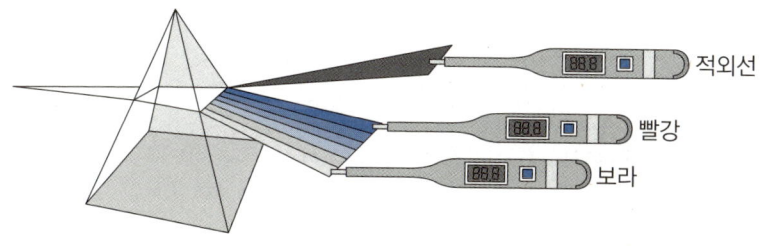

물리군 적외선은 눈에 보이지 않는데 어떻게 느낄 수 있죠?

정교수 사람이 많아지면 따뜻하지? 그건 바로 사람들이 뿜어내는 적외선 때문이야. 이 적외선의 복사로 따뜻한 열이 전달되는 거지. 사람들은 눈에 보이지 않는 적외선을 생활에 이용하기 시작했어.

적외선은 열원으로 쓸 수 있다. 예를 들어 적외선 사우나에 이용하거나, 항공기 날개에서 얼음을 제거하는 것과 같은 다른 가열 응용 분야에도 쓰인다. 또한 적외선 가열은 쿠팅 경화, 플라스틱 성형, 어닐링, 플라스틱 용접 및 인쇄 건조와 같은 산업 제조 공정에서 점점 더 대중화되고 있다.

적외선 드라이어(출처: Pittigrilli/Wikimedia Commons)

적외선을 이용하면 물체의 온도를 원격으로 결정할 수 있다. 이를 서모그래피라고 한다. 열화상 카메라는 전자기 스펙트럼의 적외선 범위(약 9,000~14,000nm 또는 9~14μm)에서 방사선을 감지하고 해당 방사선의 이미지를 생성한다. 적외선은 온도에 따라 모든 물체에서 방출된다. 그러므로 흑체복사 법칙에 의해 서모그래피를 사용하면 가시광선 여부에 관계없이 주

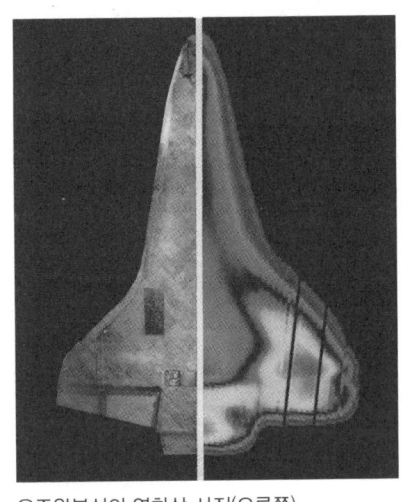

우주왕복선의 열화상 사진(오른쪽)

변 환경을 볼 수 있다. 물체에서 방출되는 방사선 양은 온도에 따라 증가하므로 서모그래피를 통해 온도 변화를 볼 수 있다.

적외선은 볼 수 있는 가시광선이 충분하지 않을 때 야간 투시 장치에 사용된다. 야간 투시 장치는 주변광 광자를 전자로 변환해 증폭한 후 가시광선으로 변환하는 과정을 통해 작동한다. 적외선 광원은 야간 투시 장치로 변환 가능한 주변광을 증강하는 데 쓰여, 실제로 가시광선 없이도 어둠 속에서 가시성을 높일 수 있다.

적외선 반사 촬영은 그림에 적용하여 비파괴적인 방식으로 기본 레이어를 드러낼 수 있다.

적외선을 이용한 야간 투시 장치(출처: www.ExtremeCCTV.com)

특히 작가의 밑그림이나 가이드로 그려진 윤곽선이 드러난다. 미술 보존가는 이 기술로 눈에 보이는 페인트의 층이 밑그림 또는 그 사이의 층과 어떻게 다른지 조사한다. 이것은 그림이 원래 예술가의 주요 버전인지 사본인지, 복원 직업에 의해 변경되었는지를 결정하는 데 매우 유용하다. 최근 적외선에 민감한 카메라의 발전으로 작가가 나중에 덧칠한 전체 그림을 발견하고 묘사할 수 있게 되었다.

모나리자의 적외선 반사 사진

물리군   자외선은 누가 발견했나요?

정교수   1801년 독일 물리학자 요한 빌헬름 리터는 가시 스펙트럼 반대편의 보라색보다 짧은 스펙트럼 빛을 관찰했어. 그는 빛에 반응하는 염화은을 바른 종이를 가지고 보라색 외부의 눈에 보이지 않는 빛을 발견했는데, 이게 바로 자외선이지.

물리군   전자레인지에서도 눈에 보이지 않는 빛이 나오죠?

정교수   맞아. 전자레인지에서는 마이크로파라는 눈에 안 보이는 전자기파가 발생해. 그럼 전자레인지가 어떻게 발명되었는지 알아볼까? 전자레인지를 발명한 사람은 스펜서(Percy Spencer, 1894~1970)야.

스펜서는 미국 메인주 하울랜드에서 태어났다. 그가 18개월 되었을 때 아버지는 세상을 떠났다. 어머니는 스펜서의 숙모와 삼촌에게 그를 맡겼다. 삼촌은 스펜서가 겨우 일곱 살이었을 때 사망했다. 스펜서는 숙모를 부양하느라 어린 나이에 돈을 벌어야 했다. 그는 12세부터 16세까지 공장에서 일했다.

18세에 스펜서는 미 해군에 입대했다. 그는 타이태닉호가 침몰했을 때 승선했던 무선통신사에 대해 알게 된 후부터 무선통신에 관심을 가졌다. 해군으로 복무하며 그는 라디오 기술을 배웠고 삼각법, 미적분학, 화학, 물리학, 야금술 등을 독학했다.

1939년까지 스펜서는 레이더 튜브 설계 분야에서 세계 최고의 전문가로 손꼽혔다. 그는 미 국방부의 계약 업체인 레이시온(Raytheon)에서 전력관 사업부장으로 일했다. 레이시온에서 근무하는 동안 스펜서는 마이크로파 발생장치인 마그네트론을 보다 효율적으로 제조하는 방법을 개발했다. 이것으로 마그네트론 하루 생산량을 100개에서 2600개로 늘렸다.

스펜서는 명성과 전문성을 모두 갖추고 있었다. 그는 레이시온이 정부 계약을 따내도록 도왔는데, MIT의 방사선 연구소를 위한 전투 레이더 장비를 개발하고 생산하는 일이었다. 이것은 제2차 세계

마그네트론(출처: HCRS Home Labor Page/ Wikimedia Commons)

대전의 연합국에게 매우 중요했으며, 전쟁 기간에 맨해튼 프로젝트 다음으로 군대의 두 번째로 높은 순위 프로젝트였다. 이 업적으로 스펜서는 미 해군으로부터 공로상을 수상했다.

한편 전자레인지는 우연히 개발된 발명품이다. 스펜서는 레이더 장비를 연구하던 중에 전자레인지를 발명했다. 그는 레이더 기계류의 동력 부품 중 하나인 마그네트론 옆에서 잠시 휴식을 취했다. 그런데 주머니에 들어 있던 초콜릿 바가 녹아버린 것을 발견하고는 깜짝 놀랐다. 호기심으로 그는 다른 여러 물체를 마그네트론 근처에 올려놓고 물체로부터 멀찌감치 서 있었다. 팝콘이 성공적으로 튀겨졌으며 다음 날 아침에는 계란이 익었다.

스펜서는 자신이 발견한 것의 가능성을 깨닫고 더욱 효율적인 음식 조리 장치를 디자인하기 시작했다. 그는 1945년에 특허를 신청했으며, 1946년경 매사추세츠주 보스턴에 있는 레스토랑에서 시제품 장치를 테스트한 후 곧바로 상업 모델을 제작하였다.

초기 전자레인지 모델은 높이가 6피트(1.8미터) 이상, 가격은 5천 달러나 되었다. 또한 마그네트론 장치를 냉각시키는 특수한 배관이 필요했기 때문에 소비자들에게 잘 판매되지 않았다. 그러나 점차 실용적으로 개선되고 가격이 하락하여 일반 소비자가 사용하기에 충분히 안전하고 신뢰할 만한 수준이 되었다.

1975년경 전자레인지의 판매액이 가스 조리 기구의 판매액을 초과했다. 전자레인지는 조리 시간이 짧아 음식이 고르게 조리되지 않는 단점에도 불구하고, 많은 사람이 꼭 필요한 가전제품으로 인식하고 있다.

초창기의 전자레인지
(출처: Acroterion/Wikimedia Commons)

두 번째 만남

레이저의 탄생

광자의 탄생 _ 빛을 이루는 입자

정교수 양자광학을 이해하려면 양자(quantum)에 대해 조금 알 필요가 있어. 양자라는 이름은 1900년 플랑크의 논문에 처음 등장했다네. 고전물리학의 입자는 연속적인 에너지를 가지는 데 반해, 양자는 불연속적인 에너지를 가지는 기묘한 입자야. 플랑크는 빛이 광자(photon)[1]라는 양자로 이루어져 있고 광자 하나가 가지는 에너지가

$$E = h\nu$$

인 것을 알아냈지. 여기서 h는 플랑크 상수이고 ν는 빛의 진동수야. 플랑크 상수는

$$h = 6.626 \times 10^{-34} (\text{J} \cdot \text{s})$$

이지. 그러니까 진동수가 ν인 광자들이 가질 수 있는 에너지는

$$h\nu, 2h\nu, 3h\nu, \cdots$$

이므로 불연속적이라네.

---

[1] 광자라는 용어는 1926년 루이스(Gilbert Lewis)가 처음 사용했다.

그 후 아인슈타인의 광전효과 논문과 콤프턴이 발표한 광자와 전자의 충돌에 관한 논문을 통해 광자가 빛을 이루는 입자라는 것이 확인되었어.

아인슈타인의 유도방출 이론 _ 레이저의 기본 원리

정교수 1913년 보어는 수소의 선스펙트럼을 연구하던 중 수소 원자 속의 전자 역시 불연속적인 에너지를 가져야 한다는 걸 깨달았지.

물리군 그렇다면 전자도 양자군요.

정교수 맞아. 보어는 전자가 원자핵 주위의 불연속적인 궤도에 존재하는 것을 알아냈어.

보어(Niels Bohr, 1885~1962, 1922년 노벨 물리학상 수상)

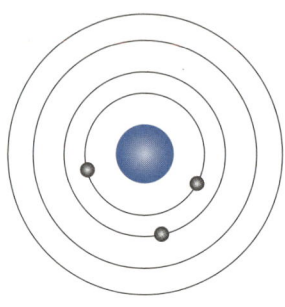

1917년 아인슈타인은 전자가 가질 수 있는 에너지와 광자 사이의 관계를 연구해 훗날 레이저의 기본 원리가 되는 유도방출에 대한 논문을 발표했다.

아인슈타인(Albert Einstein, 1879~1955, 1921년 노벨 물리학상 수상)

아인슈타인은 원자 속 전자에 허용되는 에너지 상태가 두 가지뿐인 간단한 경우를 생각했다. 전자가 가질 수 있는 두 에너지 상태 중에서 높은 에너지를 E_2, 낮은 에너지를 E_1이라고 하자. 전자가 에너지

E_1을 가질 때 전자는 에너지 E_1인 양자 상태에 있다고 말하고, 전자가 에너지 E_2를 가질 때 전자는 에너지 E_2인 양자 상태에 있다고 말한다. 전자는 두 에너지 상태 중 하나에 있게 되는데, 아인슈타인은 이 과정에서 광자의 방출과 흡수가 관련 있다고 보았다.

아인슈타인은 방출 과정에 두 종류가 있다고 생각했다. 외부 자극 없이 높은 에너지 상태의 전자가 저절로 낮은 에너지 상태로 변하면서 그 차이에 해당하는 에너지를 가진 광자가 방출되는 과정을 자발적 방출이라고 불렀다.

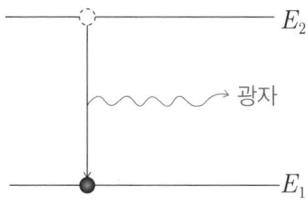

다른 경우로 외부 자극에 의해 높은 에너지 상태의 전자가 낮은 에너지 상태로 변하면서 광자가 방출될 수 있는데, 아인슈타인은 이것을 유도방출 과정이라고 불렀다. 외부에서 원자에 빛을 쪼여주면 그 빛이 외부 자극이 된다.

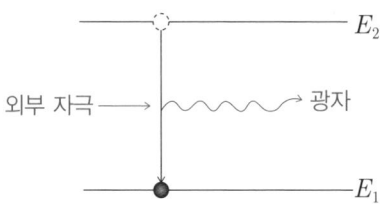

원자 속 전자는 외부의 광자를 흡수해 낮은 에너지 상태에서 높은 에너지 상태로 전이될 수 있는데, 이것은 광자의 흡수 과정이다.

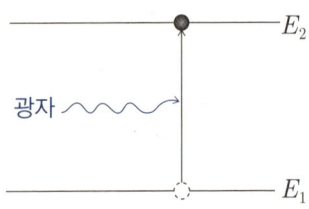

양자역학의 탄생 _ 전자라는 양자가 만족하는 물리학

정교수 자! 이번에는 양자역학의 탄생 과정을 알아볼 거야.

물리군 양자역학의 주인공은 광자인가요?

정교수 아니. 전자가 양자역학의 주인공이야. 물리학자들은 전자라는 양자가 만족하는 물리학을 만들기 시작했어. 그렇게 해서 완성된 물리학이 바로 양자역학이야.

1925년에 하이젠베르크의 불확정성원리가 발견되고, 1926년에 슈뢰딩거 방정식이 발표되었다. 이로써 사람들은 원자 속의 전자가 양자역학이라는 새로운 역학을 따르는 것을 알게 되었다.

고전역학에서 입자의 위치를 x, 운동량을 p, 에너지를 H라고 하면 질량이 m인 입자의 에너지는

$$H = \frac{1}{2m}p^2 + V(x)$$

이다. 이때 $V(x)$는 퍼텐셜에너지[2]이다.

양자역학에서는 고전역학의 위치, 운동량, 에너지가 모두 연산자로 바뀐다.

$$x \quad \longrightarrow \quad \hat{x}$$
$$p \quad \longrightarrow \quad \hat{p}$$
$$H \quad \longrightarrow \quad \hat{H}$$

하이젠베르크 및 보른과 요르단은 위치 연산자와 운동량 연산자 사이의 관계가

$$[\hat{x}, \hat{p}] = i\hbar$$

라는 것을 알아냈다. 여기서

$$\hbar = \frac{h}{2\pi}$$

이고

$$[A, B] = AB - BA$$

이다.

---

2) 위치에너지라고도 부른다.

물리군 위치 연산자와 운동량 연산자에 대하여 교환법칙이 성립하지 않는군요.

정교수 맞아. 이제 우리는 질량이 m인 전자를 양자로 취급할 거야. 그러면 에너지 연산자 $\hat{H}$는

$$\hat{H} = \frac{1}{2m}\hat{p}^2 + V(\hat{x})$$

가 돼. 물리학자들은 전자의 에너지가 불연속적이라는 사실로부터 허용된 에너지 상태를

$$E_1 < E_2 < E_3 < E_4 < \cdots$$

로 나타내고, 각 에너지 E_i에 관한 상태벡터를 $|i\rangle$로 표현했어. 그러니까 에너지 연산자는 정사각행렬로, 상태벡터는 열행렬(열벡터)로 생각하면 돼. 상태벡터를 양자 상태라고도 불러. 이때 슈뢰딩거 방정식은

$$\hat{H}|i\rangle = E_i|i\rangle \quad (i = 1, 2, 3, \cdots)$$

가 되지.

물리군 $|i\rangle$가 뭔지 잘 모르겠어요.

정교수 조금 간단한 예를 먼저 살펴볼까? 허용된 양자 상태가 두 가지뿐인 경우를 생각해 봐.

전자가 가질 수 있는 에너지가 E_1과 E_2이고

$E_1 < E_2$

라고 하자. 그리고 불연속적인 에너지 E_1, E_2에 해당하는 양자 상태를 다음과 같이 쓰자.

$|1\rangle, |2\rangle$

여기서 에너지 연산자를 양자 상태에 작용하면

$\hat{H}|1\rangle = E_1|1\rangle$

$\hat{H}|2\rangle = E_2|2\rangle$

이다. 위 식은 다음과 같이 행렬로 나타낼 수 있다.

$|1\rangle = \begin{pmatrix} 1 \\ 0 \end{pmatrix}$

$|2\rangle = \begin{pmatrix} 0 \\ 1 \end{pmatrix}$

이때 주어진 행렬의 열과 행을 바꾸고 복소수 켤레를 적용하는 것을 수반(adjoint)이라 하고, †로 표현한다.

$|1\rangle^\dagger = (1\ 0)$

$|2\rangle^\dagger = (0\ 1)$

이 경우는 복소수가 아니므로 켤레를 취하지 않았다. 예를 들어

$$|a\rangle = \begin{pmatrix} 1 \\ i \end{pmatrix}$$

이면

$$|a\rangle^\dagger = (1 \ -i)$$

가 된다. 디랙은 $|\ \rangle$의 수반을 $\langle\ |$로 나타냈다. 즉, 다음과 같다.

$$\langle 1 | = | 1 \rangle^\dagger = (1 \ 0)$$

$$\langle 2 | = | 2 \rangle^\dagger = (0 \ 1)$$

이때 $|n\rangle^\dagger = \langle n |$ 과 $|m\rangle$의 곱을

$$\langle n | m \rangle$$

으로 쓰고, 두 상태 $|n\rangle$과 $|m\rangle$의 내적이라고 한다. 여기서 $n = m$이면 $\langle n | m \rangle = 1$이고, $n \ne m$이면 $\langle n | m \rangle = 0$일 때 이를 직교정규화된 상태라고 한다. 정규화된 상태 $|n\rangle$의 크기를

$$\| |n\rangle \| = 1$$

로 표현하는데, 디랙은 이것을 다음과 같이 썼다.

$$\| |n\rangle \|^2 = \langle n | n \rangle = 1$$

앞의 예에서

$$\langle 1|1\rangle = \langle 2|2\rangle = 1$$

$$\langle 1|2\rangle = 0$$

이므로 $|1\rangle$, $|2\rangle$는 직교정규화된 상태이다. 이때 에너지 연산자는 다음과 같은 행렬로 쓸 수 있다.

$$\hat{H} = \begin{pmatrix} E_1 & 0 \\ 0 & E_2 \end{pmatrix}$$

이렇게 연산자는 행의 수와 열의 수가 같은 정사각행렬로, 양자상태는 열이 하나인 행렬로 나타낼 수 있다. 따라서 $\hat{H}|n\rangle$은 두 행렬의 곱이다. 행렬의 곱에 대한 수반 연산에는 다음 성질이 있다. 두 행렬 A, B에 대해

$$(AB)^\dagger = B^\dagger A^\dagger$$

가 성립한다.

물리군 모든 게 행렬로 표현되는군요.

정교수 물론이야. 이렇게 에너지가 두 종류만 허용되는 계에서 일반적인 상태는 허용되는 각 에너지에 대응하는 상태벡터의 중첩으로 나타낼 수 있어. 일반적인 상태를 $|\psi\rangle$라고 하면

$$|\psi\rangle = c_1|1\rangle + c_2|2\rangle$$

가 돼. 여기서 c_1, c_2는 임의의 복소수야. 이렇게 양자역학에서 전자의 상태는 복소수로 표현할 수 있지.

메이저의 발명 _ 강렬하고 정밀한 빔을 만들다

정교수 레이저는 하나의 파장을 가진 빛을 방출해. 즉, 단색광이지. 레이저에서 나온 광선을 레이저 빔이라고 하는데, 이 빔은 가늘고 잘 퍼지지 않는 특성이 있어. 하지만 레이저보다 먼저 발명된 건 메이저야. 이제 그 이야기를 해볼게.

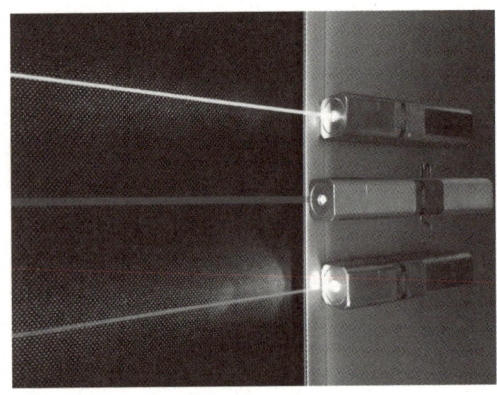

레이저포인터(출처: 彭家杰/ Wikimedia Commons)

물리군 레이저는 알겠는데 메이저는 뭐예요?

정교수 레이저(LASER)는 Light Amplification by the Stimulated Emission of Radiation의 머리글자야. 우리말로는 유도방출 광선 증폭이지. 좁은 의미의 레이저는 가시광선을 뜻하지만, 이제는 가시광선이 아닌 경우에도 레이저라고 불러.

메이저(MASER)는 Microwave Amplification by Stimulated Emission of Radiation의 머리글자야. 우리말로는 유도방출 마이크로파 증폭이지.

레이저와 메이저는 여러 과학자의 노력으로 발명되었다. 먼저 과학자 파브리칸트를 알아보자.

파브리칸트(Valentin Aleksandrovich Fabrikant, 1907~1991, 사진 출처: Biberman L.M.)

파브리칸트는 1907년 모스크바에서 태어났다. 그는 1930년 모스크바 대학의 물리 수학과를 졸업했다. 졸업 후에는 모스크바 전력 공학 연구소에 들어갔다. 1932년부터 파브리칸트는 가스 방전 광학 문제를 연구했다. 1938년 그는 유도방출의 존재를 실험으로 확인하는 데 성공했고, 유도방출을 이용해 전자기파를 증폭시킬 수 있다는 것을 알아냈다. 그 후 1947년 램(Willis E. Lamb, 1913~2008, 1955년 노벨 물리학상 수상)이 수소 스펙트럼에서 유도방출을 발견했다.

이번에는 과학자 카스틀레르를 살펴보자.

카스틀레르(Alfred Kastler, 1902~1984, 1966년 노벨 물리학상 수상)

카스틀레르는 게브빌러(Guebwiller)³에서 태어났다. 그는 1921년 파리 고등사범학교에 들어갔다. 학업을 마치고 1926년부터 뮐루즈 고등학교에서 물리학을 가르쳤다. 그 후 보르도 대학에서 1941년까지 교수로 재직하다가 파리 고등사범학교의 교수가 되었다.

---

3) 당시에는 독일제국의 도시였고 현재는 프랑스의 도시이다.

카스틀레르는 양자역학을 이용해 빛과 원자 간의 상호작용을 연구했다. 이를 통해 '광학 펌핑' 기술을 개발한 업적으로 1966년 노벨 물리학상을 수상했다.

물리군 광학 펌핑이 뭔가요?
정교수 빛을 이용해 낮은 에너지 상태의 전자를 높은 에너지 상태로 전이시키는 걸 말해.
물리군 그렇군요.
정교수 이번에는 메이저와 레이저에 대해 최초로 공개 강연을 했던 웨버를 소개할게. 그는 이 연구로 노벨상 후보로도 두 번 지명되었지만 아쉽게 노벨상을 놓쳤지.

웨버는 1919년 미국 뉴저지주 패터슨에서 태어났다. 그는 패터슨 이스트사이드 고등학교에서 기계공학을 전공했고, 1935년에 졸업 후 미국 해군사관학교에 입학했다.

제2차 세계대전 중 웨버는 미 해군 함정에서 복무했으며 중령으로 진급했다. 그는 산호해 해전에서 일본 항공모함 쇼호를 침몰시켰고 1943년 7월에는 시칠리아 침공에 참전했다.

웨버는 1943년부터 1945년까지

웨버(Joseph Weber, 1919~2000)

해군 대학원에서 전자공학을 공부했다. 이후 1948년까지는 워싱턴 D.C.에 있는 해군 선박국의 전자 대책 설계를 이끌었다. 1948년에 그는 메릴랜드 대학교 칼리지파크의 공학 교수가 되었다. 교수 재직 중 마이크로파 분광학으로 박사 과정을 밟아 1951년 미국 가톨릭 대학교에서 박사 학위를 받았다.

당시 웨버는 마이크로파 방출에 대한 아이디어를 냈다. 그는 1952년 6월에 오타와에서 열린 전자 튜브 학회에서 레이저와 메이저의 원리를 가지고 최초로 공개 강연을 했다. 이 발표가 끝나고 타운스는 웨버에게 발표한 논문 사본을 요청했다. 얼마 후 타운스는 메이저를 발명해 노벨 물리학상을 받는다. 웨버는 1962년, 1963년에 노벨 물리학상 후보에 오르지만 결국 노벨상과는 인연이 없었다.

이제 메이저 발명으로 노벨 물리학상을 수상한 세 명의 과학자를 알아보자. 먼저 타운스에 대한 이야기로 시작하겠다.

타운스는 미국 사우스캐롤라이나주 그린빌에서 태어났다. 그의 형 헨리 키스 타운스 주니어(Henry Keith Townes Jr., 1913~1990)는 맵시벌(Ichneumon wasps)에 대한 세계적인 권위자였다.

퍼먼 대학 물리학과를 졸업한 타운스는 1937년 듀크 대학에서 물리학 석사 학위를 취득했다. 그 후 캘리포니

타운스(Charles Hard Townes, 1915~2015, 1964년 노벨 물리학상 수상)

아 공과대학에서 1939년에 박사 학위를 받았다. 제2차 세계대전 중 그는 벨 연구소에서 레이더 폭격 시스템을 연구했다.

1950년 타운스는 컬럼비아 대학교의 교수가 되었다. 1951년 그는 간섭성 방사선의 강렬하고 정밀한 빔을 생성하는 새로운 방법을 고안했다. 여기에서 메이저(MASER, 방사선 유도방출에 의한 마이크로파 증폭)라는 약어를 처음 사용했다. 마이크로파 대신 눈에 보이는 빛이 나오는 경우에는 레이저라는 용어를 썼다.

1953년에 타운스는 고든(James P. Gordon), 자이거(Herbert J. Zeiger)와 함께 컬럼비아 대학교에서 최초의 암모니아 메이저를 제작했다. 이 장치로 그들은 암모니아 분자에 유도방출을 시켜 24기가헤르츠의 주파수를 가진 마이크로파 증폭에 성공했다.

최초의 암모니아 메이저와 타운스

1959년부터 1961년까지 타운스는 컬럼비아 대학교를 휴직하고 워싱턴 D.C.에 있는 국방연구원(Institute for Defense Analysis)의 부회장이자 연구 책임자로 재직했다. 그리고 1967년까지 매사추세츠 공과대학(MIT)에서 학장 겸 물리학 교수로 일했다. 1967년에 그는 버클리 대학교의 물리학 교수로 임명되어 거의 50년 동안 머물렀다.

두 번째로 소개할 과학자는 바소프이다.

바소프(Nikolay Gennadiyevich Basov, 1922~2001, 1964년 노벨 물리학상 수상)

바소프는 러시아 리페츠크주에 있는 우스만에서 태어났다. 그는 1941년 보로네시에서 학교를 마쳤다. 1943년 사관학교를 졸업하고 붉은 군대에서 복무하며 제1차 러시아-우크라이나 전쟁과 제2차 세계대전에 참여했다. 바소프는 1950년 모스크바 공학물리학 연구소(MEPhI)를 졸업한 후 이곳의 교수로 지내는 동시에 레베데프 물리학 연구소에서 연구를 진행했다.

세 번째 과학자 프로호로프에 대해 알아보자.

프로호로프(Aleksandr Mikhailovich Prokhorov, 1916~2002, 1964년 노벨 물리학상 수상)

프로호로프는 1916년 호주 퀸즐랜드주 피라몬의 러셀로드에서 태어났다. 그의 부모는 차르 정권의 탄압을 피해 러시아에서 이주한 혁명가 미하일 이바노비치 프로호로프와 마리야 이바노브나였다.

1923년 10월 혁명과 러시아 내전이 끝나고 프로호로프 가족은 러시아로 돌아왔다. 1934년 프로호로프는 물리학을 공부하기 위해 상트페테르부르크 주립대학교에 입학했다. 그리고 1939년에 우수한 성적으로 졸업한 후 모스크바로 이주하여 레베데프 물리학 연구소에서 일했다. 그곳에서 그는 대기 전리층에서의 전파를 연구했다.

제2차 세계대전이 발발한 1941년 6월, 프로호로프는 붉은 군대에 입대했다. 그는 보병으로 싸우다 1944년 제대했으며, 1946년에 세 개의 훈장을 받았다. 레베데프 물리학 연구소로 돌아온 뒤 1946년에 튜브 발진기의 주파수 안정화 이론으로 박사 학위를 받았다.

1947년 프로호로프는 싱크로트론(synchrotron)이라고 불리는 순환 입자 가속기에서 궤도를 도는 전자에 의해 방출되는 간섭성 방사선을 연구하기 시작했다. 그는 이 방출이 대부분 마이크로파임을 알아냈다. 1950년까지 프로호로프는 레베데프 물리학 연구소의 부국장으로 일했다.

1950년대 초에 바소프와 프로호로프는 분자 발진기를 만들기 위한 이론적 근거를 개발하고 암모니아를 기반으로 이러한 발진기를 구상했는데, 이것이 바로 메이저였다.

아내들과 함께. 바소프와 프로호로프

메이저를 만든 공로로 타운스와 바소프, 프로호로프는 1964년 노벨 물리학상을 수상했다.

레이저의 발명 _ 증폭된 가시광선의 빔

정교수 이제 레이저 발명에 관한 이야기를 하려고 해.

물리군 레이저의 발명으로 노벨상은 받은 과학자는 누구인가요?

정교수 메이저와 레이저의 원리가 거의 같아서 레이저의 발명으로 노벨상을 수상한 과학자는 없어.

물리군 그렇군요.

정교수 메이저에서 나오는 증폭된 마이크로파는 눈에 보이지 않아. 그래서 과학자들은 증폭된 가시광선의 빔을 만드는 일에 관심을 돌렸지. 이것이 바로 레이저야.

레이저 발명과 관련된 두 명의 과학자를 알아보자. 먼저 숄로를 소개하겠다.

숄로는 미국 뉴욕주 마운트버넌에서 태어났다. 그의 어머니는 캐나다 출신이었고, 아버지는 당시 러시아제

숄로(Arthur Leonard Schawlow, 1921~1999, 1981년 노벨 물리학상 수상, 사진 출처: Jose Mercado/Stanford News Service)

국의 리가(현재 라트비아의 수도) 출신인 유대인 이민자였다.

숄로는 세 살 때 캐나다 온타리오주 토론토로 이사했다. 그는 16세에 본로드 고등학교(VRA)를 졸업하고 토론토 대학교 빅토리아 칼리지에 입학했다. 제2차 세계대전으로 잠시 학업을 중단했던 그는 전쟁이 끝난 후 박사 과정을 이수했다. 그리고 1949년 가을에 컬럼비아 대학교 물리학과에서 타운스와 함께 박사 후 과정을 밟았다.

두 번째로 우리가 알아야 할 과학자는 굴드이다.

굴드(Richard Gordon Gould, 1920~2005, 사진 출처: Union College)

미국 뉴욕에서 태어난 굴드는 세 아들 중 맏이였다. 그의 아버지는 뉴욕시에 있는 스콜라스틱 매거진 출판사의 창립 편집자였다. 굴드는 뉴욕 교외의 작은 마을인 스카스데일에서 자랐고 스카스데일 고등학교에 다녔다. 그는 유니언 칼리지에서 물리학 학사 학위를 받고 예일 대학교에서 광학 및 분광학을 전공으로 석사 학위를 받았다.

1944년 3월부터 1945년 1월까지 굴드는 맨해튼 프로젝트에 참여했지만, 공산주의 정치 협회 회원 활동 때문에 해고되었다. 1949년에는 컬럼비아 대학교에서 광학 및 마이크로파 분광법 박사 과정을 밟았다.

물리군 누가 레이저를 발명한 거죠?

정교수 이제 그 이야기를 해볼게. 1957년까지 타운스를 포함한 많은 과학자는 가시광선을 메이저처럼 증폭시키는 방법을 찾고 있었어. 그해 11월 굴드는 자신의 실험 결과를 적은 〈레이저의 타당성에 대한 몇 가지 대략적인 계산: 방사선 유도방출에 의한 광 증폭〉이라는 제목의 노트를 만들었다네. 이 노트에는 실행 가능한 레이저를 만드는 방법이 적혀 있었지.

물리군 굴드가 최초로 레이저를 발명한 셈이군요.

정교수 레이저라는 이름을 처음 사용한 사람이 바로 굴드야. 하지만 그는 이 노트를 논문으로 발표하지 않았어.

물리군 왜죠?

정교수 굴드는 이 노트 내용을 토대로 레이저에 대한 특허를 내고 싶었거든. 그는 대학에서 레이저를 개발하면 특허권이 온전히 자신에게 주어지지 않을 것을 염려했어. 그래서 박사 학위를 포기하고 민간 연구 회사인 TRG(Technical Research Group)에 입사했지.

굴드가 레이저에 관한 실험 노트를 완성하고 석 달 뒤, 1958년 숄로와 타운스는 레이저 이론을 만들기 위해 의기투합해 메이저와 레이저의 설계 및 작동 이론을 발표했다네.

물리군 논문으로는 숄로와 타운스가 먼저 발표했군요.

정교수 맞아. 굴드는 TRG에서 세계 최초로 작동 레이저를 만들어 특허를 가지고자 했지. 당시 TRG는 정부로부터 연구비를 지원받고 있었는데, 공산주의 활동 이력 탓에 굴드는 레이저 개발 프로젝트에 직접 기여할 수 없었

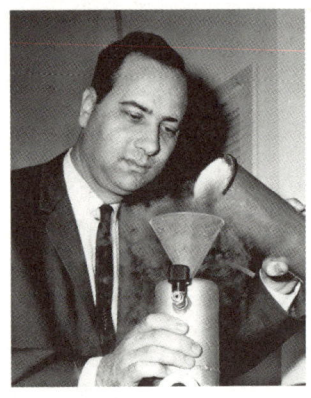

메이먼(Theodore Harold Maiman, 1927~2007)

어. 그사이에 최초의 작동 레이저가 휴스 연구소(HRL)의 메이먼에 의해 세상에 나타난 거야.

이 레이저는 루비를 사용했기 때문에 루비 레이저라고 부르는데, 파장이 694nm인 붉은색의 빔이 발생해.

세계 최초의 레이저인 메이먼의 루비 레이저(출처: Theodore and Kathleen Maiman)

물리군   굴드는 매우 아쉬워했겠네요.

정교수   물론이야. 루비 레이저 다음으로 발명된 것은 헬륨-네온 레이저로 붉은색 부분인 632.8nm의 파장에서 작동해. 이 레이저는 1962년에 벨 전화 연구소에서 개발되었지.

헬륨-네온 레이저(출처: Tommy Markstein/Wikimedia Commons)

또 다른 유명한 레이저로 이산화탄소 레이저(CO_2 레이저)가 있어. 1964년에 벨 연구소에서 발명되었지. 이 레이저는 적외선 빔을 생성해.

이산화탄소 레이저

레이저 응용 _천문학에서 의학까지

물리군 레이저는 어디에 사용되나요?

정교수 레이저가 쓰이는 곳은 너무너무 많아. 하나씩 살펴볼까?

MIT 전기공학과 교수인 스멀린(Louis Smullin)은 루비 레이저가 발명된 후, 대기 물리학자 조르조 피오코(Giorgio Fiocco)와 함께 1962년 5월 9일부터 11일까지 레이저 광선의 펄스를 달로 전송했다. 이를 가지고 새로운 정확도로 달까지의 거리를 측정했다.

피오코(Giorgio Fiocco, 1931~2012, 사진 출처: Piero Mazzinghi/Wikimedia Commons)

또 다른 응용으로 라이다가 있다. 이것은 레이저를 물체 또는 표면에 쏘여 반사된 빛이 수신기로 돌아가는 시간을 측정해 주위를 스캔하는 장치이다. 라이다는 측량, 측지학, 기하학, 고고학, 지리학, 지질학, 지형학, 지진학, 임업, 대기 물리학 등에 사용된다.

라이다(출처: David Monniaux/Wikimedia Commons)

한편 레이저를 이용한 절단 기술을 레이저 절단이라고 부른다. 레이저 절단은 고출력 레이저의 출력으로 작동한다. 절단면에 집중된 레이저 빔은 재료가 녹거나 타거나 기화하게 하며 가장자리에 고품질 표면 마감을 남긴다.

레이저 절단기(출처: Zgyricky/Wikimedia Commons)

레이저는 병원에서도 사용된다. 레이저가 발명된 지 불과 1년 후인 1961년, 찰스 캠벨(Charles J. Campbell)은 루비 레이저를 가지고 단 한 번의 펄스로 혈관종성 망막 종양을 파괴하는 데 성공했다. 1963년 리언 골드먼(Leon Goldman)은 루비 레이저를 사용하여 색소가 있는 피부 세포를 치료했다.

색소 레이저(출처: Han-Kwang/ Wikimedia Commons)

세 번째 만남

광통신의 역사

그레이엄 벨 _ 광통신을 발명하다

정교수 이제 빛을 이용한 통신의 역사를 알아볼게. 이러한 통신을 광통신이라고 하는데 최초의 현대적 광통신 기술을 발명한 사람은 그레이엄 벨이야.

벨은 1847년 스코틀랜드 에든버러에서 태어났다. 그의 아버지는 음성학자인 알렉산더 멜빌 벨이다.

벨(Alexander Graham Bell, 1847~1922)

벨의 가장 친한 친구는 이웃에 살던 벤 허드먼으로, 그의 아버지 존 허드먼(John Herdman)은 제분소를 운영했다. 벨은 어릴 때부터 발명에 소질을 보였다. 12세 때 회전하는 패들과 네일 브러시 세트를 결합해 간단한 껍질 제거 기계를 만들었고, 이것은 제분소에서 사용되었다. 그 대가로 존 허드먼은 두 아이에게 발명 실험을 할 작은 작업장을 만들어 주었다.

알렉산더 멜빌 벨(Alexander Melville Bell, 1819~1905)

감수성이 예민하고 예술, 시, 음악에 재능이 있던 벨은 정식 교육을 받지 않고도 피아노를 칠 수 있었다. 그는 성대모사도 잘해서 가족과

손님들을 즐겁게 했다. 그가 12세가 되던 해부터 어머니가 청력을 잃기 시작하자 벨은 수화 언어를 배웠고 음향학을 공부했다.

어렸을 때 벨은 형들과 마찬가지로 집에서 아버지에게 교육을 받았다. 11세 때 그는 에든버러의 왕립 고등학교에 등록했다. 그의 주요 관심 분야는 과학, 특히 생물학이었다. 하지만 의무교육 과정이 마음에 들지 않아 15세에 이 학교를 중퇴했다. 16세에 벨은 스코틀랜드 머리의 엘긴에 있는 웨스턴 하우스 아카데미에서 웅변과 음악을 가르쳤다.

1865년 벨의 가족은 런던으로 이사했다. 벨은 1868년 6월 유니버시티 칼리지 런던의 입학시험에 합격하여 가을에 입학했다. 그러나 1870년에 가족이 캐나다로 이주하면서 학업을 중단했다. 벨의 가족은 캐나다 온타리오주 브랜트퍼드에 정착했다.

캐나다에서의 벨의 집(출처: Harry Zilber/Wikimedia Commons)

보스턴 청각 장애인 학교의 교장 세라 풀러는 벨의 아버지에게 교사들을 위한 청각 장애인용 가시 언어 교육을 의뢰했다. 벨의 아버지는 그 제안을 거절하는 대신 아들인 벨을 보내겠다고 했다. 1871년 4월, 보스턴으로 간 벨은 교사들을 성공적으로 훈련했다. 그는 코네티컷주 하트퍼드에 있는 미국 청각 장애인 정신병원과 매사추세츠주 노샘프턴에 있는 클라크 청각 장애인 학교로부터 프로그램을 반복해 달라는 요청을 받았다.

1871년 보스턴 청각 장애인 학교에서 교사들에게 교육법을 제공했던 벨(계단 맨 위 오른쪽)

해외에서 6개월을 지내고 브랜트퍼드로 돌아온 벨은 전신기 실험을 계속했다. 그는 단일 와이어를 통해 메시지를 보낼 수 있는 송신기와 수신기를 연구했다.

1872년 10월, 벨은 보스턴에 음성 생리학 및 언어 역학 학교를 열었다. 이곳에는 많은 청각 장애인 학생이 모여들었다. 벨의 첫 학급에는 30명의 학생이 있었다. 헬렌 켈러도 그의 제자였는데, 그녀는 보지도 듣지도 말하지도 못하던 어린 시절에 그를 찾아왔다.

1888년 헬렌 켈러(왼쪽)와 설리번

1872년 벨은 보스턴 대학교의 음성 생리학 교수가 되었다. 1874년에 이르러 그는 고조파 전신기 연구를 시작했다. 당시 전신기는 수신에 국한되어 있었다. 즉, 전신은 모스부호로 한 번에 하나의 메시지를 보내는 방식이었다.

1874년 전신 메시지 트래픽은 급속히 확장되고 있었다. 웨스턴 유니언 회장 윌리엄 오턴(William Orton)의 말을 빌리자면 '상업의 신경계'가 되었다. 오턴은 발명가인 토머스 에디슨(Thomas Edison), 엘리샤 그레이(Elisha Gray)와 계약을 맺고, 새로운 전선을 건설하는 데 드는 막대한 비용을 줄이기 위해 각 전신선에 여러 개의 전신 메시지를 보내는 방법을 찾았다.

캐나다에 오기 전에도 벨은 잘 알려진 음악 현상을 가지고 여러 개의 전신 메시지를 동시에 전송하는 아이디어에 흥미를 느꼈다. 그는 동일한 소리굽쇠를 양쪽 끝에서 서로 다른 주파수로 조정하여 수신과 송신이 가능한 전신기를 만들었다.

1874년 10월, 벨은 여러 개의 메시지를 동시에 전송하는 데 성공했다. 그는 미래의 장인인 가디너 그린 허버드(Gardiner Greene Hubbard)에게 이 일을 알렸다. 허버드는 벨에게 필요한 재정을 지원했다. 벨은 젊은 전기 기술자 토머스 왓슨(Thomas Watson)을 고용했고, 두 사람은 함께 음성을 전기적으로 전송하는 장치를 연구했다.

1875년 6월 2일, 벨과 왓슨은 다중 전신기 실험을 하고 있었다. 옆방에서 왓슨이 금속 리드(reed)를 수리하는 동안, 벨에게 갑자기 전기의 흐름을 통해 왓슨의 소리가 뚜렷하게 들렸다. 이것이 바로 전화의 발명이었다. 벨은 그로부터 채 1년이 지나기도 전에 전화기에 대한 특허를 등록했다.

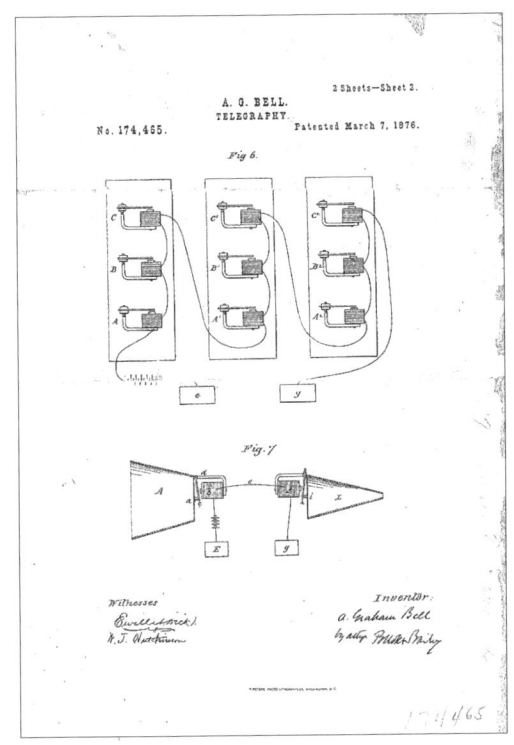

1876년 3월 7일 벨의 전화 특허 그림(출처: Alexander Graham Bell)

벨과 허버드, 샌더스, 왓슨은 1877년 7월 9일에 벨 전화 회사를 설립했다. 이틀 후 벨은 매사추세츠주 케임브리지의 허버드 저택에서 메이블 허버드(1857~1923)와 결혼했다. 벨 부부는 유럽으로 일 년간 신혼여행을 떠났다. 전화는 즉각적인 성공을 거둔 것처럼 보였지만 처음에는 수익성이 있는 사업이 아니었다. 1897년 이후까지 벨의 주요 수입원은 강의였다.

벨과 그의 아내와 자녀들

 브랜트퍼드에서 실험을 계속하면서 벨은 작동하는 전화기 모델을 만들었다. 1876년 8월 3일, 브랜트퍼드의 전신국에서 벨은 4마일(6킬로미터) 떨어진 마운트플레전트 마을로 전화를 걸었다. 8월 10일에는 브랜트퍼드와 온타리오주, 파리 사이의 8마일(13킬로미터) 떨어진 전신선을 통해 전화를 작동하는 데 성공했다. 그해 10월 9일, 케임브리지와 보스턴(약 2.5마일) 사이에 회선을 통한 최초의 쌍방향 전화 통화가 이루어졌다.

1892년 뉴욕에서 시카고까지 장거리 전화를 거는 벨

포토폰의 발명 _ 거울에 반사되는 빛으로

정교수 광통신은 빛을 이용해 정보를 먼 거리로 전달하는 방법이야. 광통신 체계는 정보를 광신호로 만드는 송신기, 신호를 목적지로 운반하는 채널, 광신호를 다시 정보로 만드는 수신기로 구성되지.

광통신의 역사는 길다. 가장 처음 등장한 광통신은 손에 횃불을 들고 흔드는 것이었다. 기원전 35만 년경 베이징 사람들은 연기를 가지고 통신하였고, 이 방법은 현재 바티칸에서 교황을 선출할 때 쓰인다.

이와 같은 방법으로 고대 그리스는 트로이가 멸망한 사실을 500킬로미터가 넘는 거리에서 알 수 있었다.

이처럼 광통신은 사전에 약속한 대로 먼 거리에 정보를 전달하기 위해 활용되었다. 봉화, 등대, 깃발 등 시각적인 기술들은 광통신의 초기 형태이다. 현재도 항해나 항공 등 먼 거리에서 위험을 알리거나 신호를 전송하는 데 사용된다.

오늘날 다양한 전자 시스템은 광신호를 송신 및 수신할 수 있다. 광섬유 통신은 위성에 의해 전달되지 않는 전자 통신에 쓰인다. 선이 없는 자유 공간 광통신 또한 다양한 분야에서 사용되고 있다. 자유 공간 광통신은 광섬유 통신과 반대로 물리적 연결이 실용적이지 않은 곳에서 유용하다. 배치가 손쉽고 비용이 저렴한 장점이 있지만, 날씨와 대기의 흡수에 영향을 받는 단점이 있다.

현대적 광통신 기술은 1880년 최초의 상업용 전화기 발명가인 알렉산더 그레이엄 벨과 테인터가 최초의 전자 장치인 광전화기(포토폰)를 발명하면서 시작되었다. 포토폰은 빛을 통해 음성을 전송하는 장치이다.

테인터는 미국 매사추세츠주 워터타운에서 태어났다. 그는 공립학교에 다녔지만 대부분 독학으로 지역 도서관에서 기술 서적을 읽으면서 공부했다.

테인터(Charles Sumner Tainter, 1854~1940)

1870년부터 테인터는 보스턴의 전기 기기 회사에서 일했다. 1873년에 그는 매사추세츠주 케임브리지포트의 망원경과 광학 기기 제조 업체인 앨번 클라크 앤드 선스(Alvan Clark and Sons)에 취직했다. 이 회사는 1874년 12월 8일에 금성의 통과를 관측하는 미국 탐험대를 위한 장비를 제작하는 계약을 체결했다. 테인터는 원정대의 일원으로 임명되어 뉴질랜드로 가서 이동 경로를 기록하고, 세계 일주를 한 뒤 워싱턴 D.C.로 돌아왔다.

1878년 테인터는 매사추세츠주 케임브리지포트에 과학 기기 생산을 위한 가게를 열었다. 그는 거기서 알렉산더 그레이엄 벨을 알게 되었다. 1년 후 벨은 테인터를 워싱턴 D.C.에 있는 볼타 연구소로 불렀고, 테인터는 그곳에서 몇 년간 일했다. 이 기간에 그는 벨과 함께 몇 가지 발명품을 만들었는데, 그중에는 포토폰과 축음기가 있다.

1899년의 축음기(출처: Norman Bruderhofer/Wikimedia Commons)

이제 두 사람이 발명한 포토폰에 대해 알아보자. 1880년 2월 19일, 벨과 테인터는 빛을 이용한 무선전화인 포토폰을 발명했다. 그해 6월 3일, 벨은 포토폰을 이용해 프랭클린 학교 지붕에서 213미터 떨어진 자신의 실험실까지 통신을 하는 데 성공했다.

포토폰

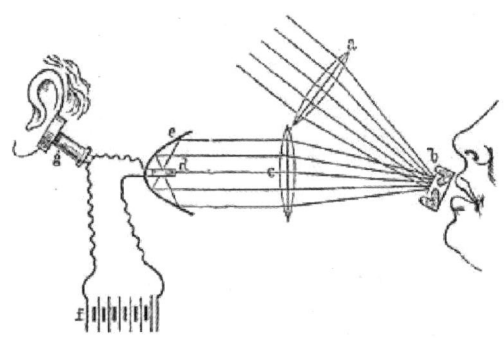

벨의 1880년 논문 중 하나에서 발췌한 포토폰에 대한 다이어그램

1880년 벨과 테인터의 광통신 시스템의 절반인 포토폰 수신기 및 헤드셋

포토폰의 송신기에는 말하는 튜브 끝에 매우 얇은 거울이 달려 있었다. 이 거울은 햇빛을 반사하는 역할을 했으며, 사용자가 말을 하면 소리의 진동이 거울에 전달되었다. 소리가 전달되면 거울은 그 진동에 따라 미세하게 움직였고, 볼록해졌다가 오목해지는 모양으로 떨렸다. 이로 인해 거울에서 반사되는 햇빛의 양이 계속해서 달라졌고, 그 결과 빛의 세기가 소리에 따라 변조되었다. 이 변조된 빛은 먼 곳에 있는 수신기로 전달되었다. 포토폰의 수신기는 셀레늄 셀이 초점에 위치한 포물선 거울로 구성되어 있다. 이 장치는 반사된 빛을 집중시켜 셀레늄 셀에 도달하게 하며, 셀레늄의 광전 특성을 이용해 빛의 세기 변화에 따라 전기적 변화를 유도한다. 그 결과 다시 음성신호로 재구성할 수 있다.

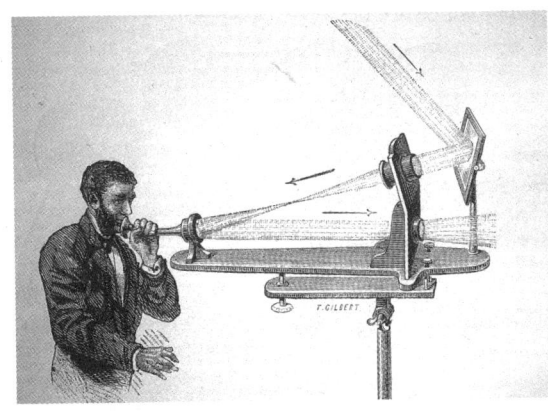

변조 전후 반사된 태양광의 경로를 보여주는 포토폰 송신기 그림

변조된 빛을 소리로 변환하는 과정과 그 전원(P)을 묘사한 포토폰 수신기 그림

1901년 독일의 물리학자 에른스트 루머(Ernst Ruhmer)는 셀레늄 셀을 개선하여 포토폰의 유효 거리를 최대 15킬로미터까지 확장했다. 하지만 루머의 포토폰은 안정적이지 않았다. 이후 독일의 지멘스-할스케(Siemens & Halske)사는 상용화를 목적으로 보다 안정적인 성능을 갖춘 모델을 개발했으며, 유효 거리는 8킬로미터 수준이었다. 1935년부터는 칼 자이스(Carl Zeiss)사가 독일 육군 전차부대를 위해 야간용 적외선 포토폰을 생산했으며, 텅스텐램프와 적외선 필터, 황화납(PbS) 검출기 및 증폭기를 사용하여 14킬로미터 거리의 통신이 가능했다.

광섬유의 발견 _ 유리 막대에 갇힌 빛

정교수 이제 광섬유를 이용한 광통신의 역사를 살펴볼까? 먼저 스위스의 물리학자 콜라동을 소개할게.

　스위스 제네바에서 태어난 콜라동은 원래 법학을 공부했다. 하지만 과학을 좋아해서 앙페르와 푸리에의 실험실에서 일했다. 그는 친구 스튀름(Charles François Sturm)과 함께 음

콜라동(Jean-Daniel Colladon, 1802~1893)

속을 측정하고 물 분사를 연구했다. 그는 청중들에게 상자에 난 구멍으로 분사되는 물줄기를 보여주기 위해 물줄기 속에 빛이 갇히는 방법을 고안했다. 이로써 청중들은 휘어지는 빛의 모양을 통해 물줄기의 궤적을 볼 수 있었다. 콜라동은 이 실험 내용을 1842년 프랑스 과학 아카데미 저널인 《Comptes Rendus》에 발표했다.

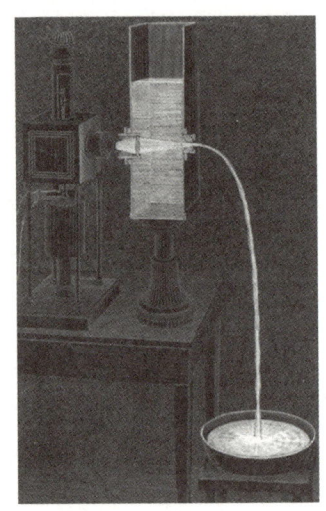

콜라동의 '빛 분수' 실험

물리군 어떻게 빛이 물줄기 안에 갇히나요?

정교수 빛이 물에서 공기로 굴절될 때, 어떤 특정한 각도보다 큰 각으로 입사하면 공기로 전혀 굴절되지 않고 물로 반사만 되거든. 이 현상을 전반사라 하고 이 특정한 각도를 임계각이라고 불러.

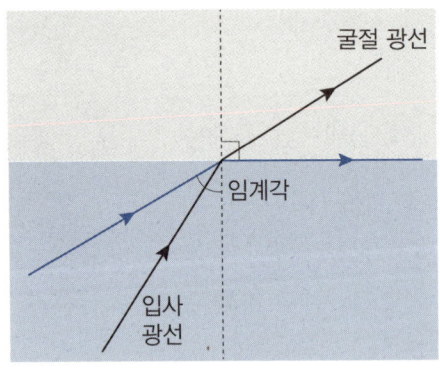

그 후 사람들은 물줄기 대신 가느다란 유리 막대에 빛이 갇히게 하여, 이것을 구부려 사람의 몸속을 조사했다. 이렇게 유리 막대나 플라스틱 막대에 빛이 갇혀 있는 형태를 광섬유라고 부른다. 이 용어는 1960년 미국의 과학자 카파니(Narinder Singh Kapany)가 처음 썼다. 광섬유는 유리나 플라스틱의 긴 섬유 막대를 통해 빛을 진행시킨다. 광섬유 케이블로 음성, 이미지 및 기타 데이터를 빛의 속도에 가깝게 전송할 수 있다.

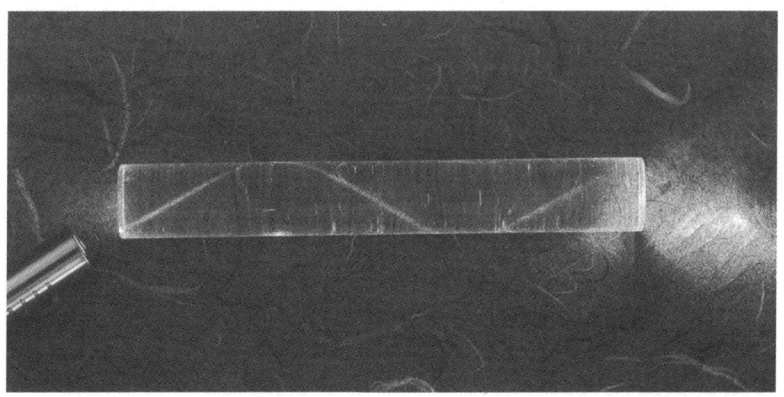

광섬유 내부에서 반사되는 레이저(출처: Timwether/Wikimedia Commons)

1888년 오스트리아 빈의 의사 로스와 로이스는 구부러진 유리 막대에 빛을 가두어 환자의 체강을 비추는 데 성공했다. 이것은 최초의 광섬유라고 볼 수 있다. 1895년 프랑스 엔지니어 앙리 생르네(Henry Saint-Rene)는 초창기 텔레비전에서 이미지를 밝게 표현하기 위해 구부러진 유리 막대로 광섬유를 만들었다.

1920년대에 영국의 베어드(John Logie Baird)와 미국의 한셀(Clarence W. Hansell)은 투명 막대 배열을 사용하여 TV용 이미지를 전송했다. 1930년대에 독일 의대생 하인리히 람(Heinrich Lamm)은 광섬유 다발로 이미지를 전송할 수 있음을 알아내고, 이를 이용해 사람의 몸속을 들여다보는 데 성공했다.

1954년 말 임페리얼 칼리지 런던의 홉킨스(Harold Hopkins)와 카파니는 10,000개 이상의 섬유로 이미지 전송 번들을 만들었다. 이후 섬유 수천 개를 결합한 길이 75센티미터의 번들로 이미지를 전송했다. 최초의 실용적인 광섬유 위내시경은 미시간 대학 연구원인 허셔위츠(Basil Hirschowitz), 피터스(C. Wilbur Peters) 및 커티스(Lawrence E. Curtiss)가 발명했다.

광섬유(출처: Ximeg/Wikimedia Commons)

저손실 광섬유의 발명과 광통신 _ 더 효율적으로 많은 정보를

정교수 이번에는 저손실 광섬유를 발명한 가오에 대해 알아볼게.

가오(Charles Kuen Kao, 1933~2018, 2009년 노벨 물리학상 수상, 사진 출처: David Dobkin/Wikimedia Commons)

가오는 1933년 중국 상하이에서 태어났다. 그는 집에서 형과 함께 가정교사로부터 중국 고전을 배웠다. 또한 상하이 국제학교에서 영어와 프랑스어를 공부했다.

공산주의 혁명 이후 1949년에 가오의 가족은 홍콩에 정착했다. 그들은 상하이 이민자들이 사는 지역인 노스포인트 가장자리에 있는 라우신 거리에 살았다. 가오는 홍콩에 있는 동안 세인트 조지프 칼리지에서 5년간 공부하고 1952년에 졸업했다.

세인트 조지프 칼리지

1953년 가오는 런던으로 건너가 중등학교에서 공부를 계속했고, 1955년에 A-LEVEL을 취득했다. 그는 나중에 울리치 공과대학(현재 그리니치 대학교)에 입학하여 전기공학 학사 학위를 받았다. 1965년에는 런던 대학교에서 전기공학 박사 학위를 따냈다.

물리군 가오는 어떤 업적으로 노벨상을 받았나요?

정교수 초기 광섬유는 에너지 손실이 너무 커서 장거리 통신으로 사용하기 곤란했어. 광섬유 1킬로미터당 손실이 무려 1000데시벨 정도였으니까 말이야.

물리군 광섬유에서 에너지 손실은 어떻게 발생하는 건가요?

정교수 반사 과정에서 생기거나 불순물과 충돌해서 생기기도 하지.

광섬유 케이블이 구부러져서 생길 수도 있어.

물리군 케이블이 구부러지면 왜 손실이 일어나죠?

정교수 케이블이 구부러지지 않았을 때, 임계각 이상으로 입사한 빛은 전반사로 케이블 안에 갇히게 돼.

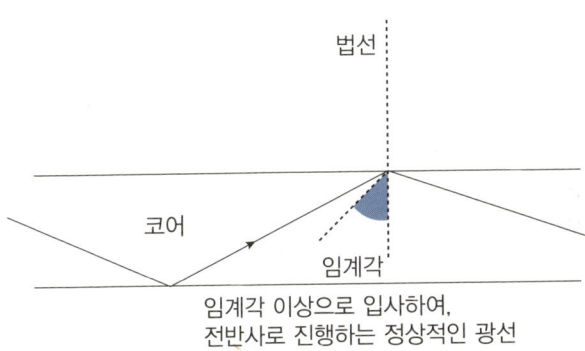

임계각 이상으로 입사하여,
전반사로 진행하는 정상적인 광선

하지만 케이블이 구부러지면 법선이 새롭게 정의되고, 이 빛이 임계각보다 작아져서 공기로 굴절되어 빠져나갈 수 있거든.

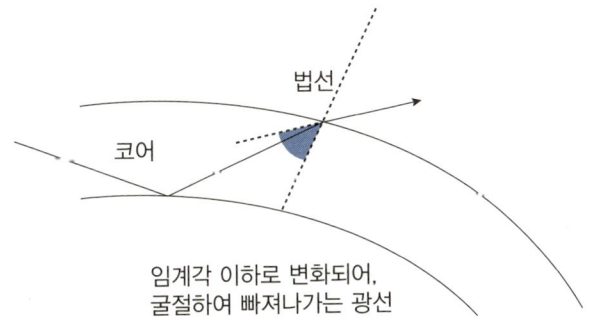

임계각 이하로 변화되어,
굴절하여 빠져나가는 광선

이러한 문제를 해결하기 위해 클래딩(Cladding)이라는 아이디어가 등장했어. 클래딩은 코어(Core)보다 굴절률이 낮은 것을 사용하여 광섬유의 에너지 손실을 줄이는 역할을 하지. 1951년 덴마크의 물리학자인 홀게르 묄레르(Holger Moeller)는 굴절률이 낮은 재료로 유리나 플라스틱 광섬유를 코팅할 것을 제안했다네. 그 후 판힐(Abraham Van Heel)은 클래딩이 유리로 된 광섬유를 만들었어.

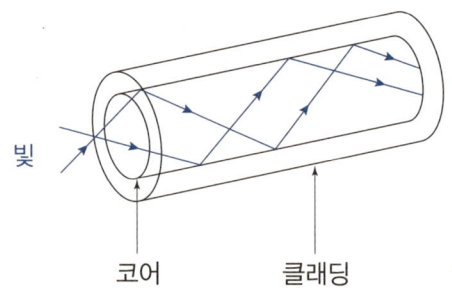

물리군 가오가 한 일은 뭔가요?

정교수 1966년 영국 스탠더드 통신연구소의 가오와 조지 호컴은 석영 유리의 불순물을 제거하면 킬로미터당 20데시벨의 손실이 발생함을 알아냈어. 이들은 석영 유리의 굴절률을 제어하고 불순물을 제거한 광섬유를 파이프 형태로 만드는 기술을 개발했지.

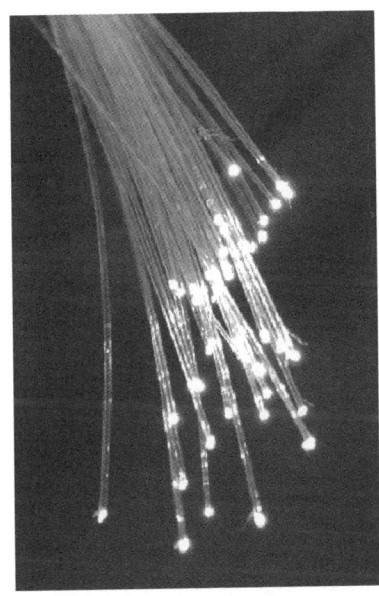

광섬유 가닥
(출처: BigRiz/Wikimedia Commons)

1970년이 되어 코닝 글라스 웍스(Corning Glass Works)의 로버트 모러(Robert Maurer), 도널드 켁(Donald Keck), 피터 슐츠(Peter Schultz), 프랭크 지마르(Frank Zimar)는 실리카 유리에 타이타늄을 도핑하여 킬로미터당 손실이 20데시벨 미만인 광섬유를 만들었다.

1970년대 후반과 1980년대에 전화 회사들은 통신망에 광섬유를 광범위하게 사용했다. 1980년대 중반에 스프린트(Sprint)는 전미 최초로 100% 디지털 광섬유 네트워크를 설립했다. 1986년 사우샘프턴 대학의 데이비드 페인(David Payne)과 벨 연구소의 에마뉘엘 데수르비어(Emmanuel Desurvire)가 개발한 어븀 도핑 광섬유 증폭기는 장거리 시스템 비용을 절감했다. 광섬유를 이용한 최초의 대서양 횡

단 전화 케이블은 1988년에 데수르비어의 레이저 증폭 기술을 활용하여 운영되었다.

1991년에 데수르비어와 페인은 광섬유 케이블 자체에 내장된 광 증폭기를 시연했다. 이 모든 광학 시스템은 전자 증폭기가 있는 케이블보다 100배 많은 정보를 지원할 수 있었다. 1991년에는 광자 결정 섬유가 등장했다. 회절을 통해 빛을 유도하는 이 광섬유는 기존 광섬유보다 더 효율적으로 전력을 전달할 수 있었다.

광 증폭기를 사용하는 전광 광섬유 케이블인 TPC-5는 1996년에 태평양을 가로질러 설치되었다. 이듬해 FLAG(Fiber Optic Link Around the Globe)는 가장 긴 단일 케이블 네트워크가 되었다. 차세대 인터넷 애플리케이션은 이를 기반으로 삼았다.

오늘날 기술 산업의 다양한 응용 분야에서 광섬유를 찾을 수 있다. 군사, 의료, 통신, 데이터 스토리지, 네트워크 및 방송 산업은 모두 이 다용도 광섬유를 활용한다.

네 번째 만남

양자광학의 탄생

광자의 생성·소멸 연산자 _ 광자 한 개를 어디로 보낼까?

정교수 이제 광자의 생성·소멸 연산자를 도입한 물리학자 디랙의 논문을 살펴볼게. 디랙은 우리가 관측할 수 있는 공간에 광자가 한 개도 없는 상태를 진공 상태라고 불렀어.

디랙(Paul Dirac, 1902~1984, 1933년 노벨 물리학상 수상)

디랙은 우리가 관측할 수 있는 공간에 광자의 개수가 n개인 상태를 나타내는 상태 벡터를

$$|n\rangle$$

이라고 쓰기로 했다. 그러니까 진공 상태를 나타내는 상태 벡터는

$$|0\rangle$$

으로 쓰면 될 것이다.

여기서 0은 광자가 0개라는 뜻이다. 이 상태는 정규화된 상태로 약속한다. 즉, 이 상태 벡터의 크기는 1이다. 이것을 수식으로 쓰면 다음과 같다.

$$\langle 0|0 \rangle = \||0\rangle\|^2 = 1$$

일반적으로 광자수가 n개인 상태도 정규화된 상태로 약속한다. 따라서

$$\langle n|n \rangle = \||n\rangle\|^2 = 1 \qquad (4\text{-}1\text{-}1)$$

이 된다.

디랙은 우리가 관측할 수 있는 공간에서 광자수를 하나 감소시키는 연산자를 $\hat{a}$로 나타내고 소멸 연산자라고 불렀고, 광자수를 하나 증가시키는 연산자를 $\hat{a}^\dagger$로 나타내고 생성 연산자라고 불렀다. 여기서 $\hat{a}^\dagger$는 $\hat{a}$의 '수반'이다. 그러니까

$$(\hat{a}^\dagger)^\dagger = \hat{a}$$

라고 할 수 있다.

물리군 생성 연산자는 숨어 있는 공간 속의 광자 한 개를 관찰 가능한 공간으로 보내고, 소멸 연산자는 관찰 가능한 공간에 있는 광자 한 개를 숨어 있는 공간으로 보내는 역할을 하는군요.

정교수 맞아. 디랙은 광자와 같은 보존은 다음 관계를 만족한다고 생각했지.

$$\hat{a}\hat{a}^\dagger - \hat{a}^\dagger\hat{a} = I \qquad (4\text{-}1\text{-}2)$$

여기서 I는 항등 연산자이다. 그러니까

$$I|n\rangle = |n\rangle \qquad (4\text{-}1\text{-}3)$$

이다.

즉, 진공 상태에 $\hat{a}^\dagger$를 작용하면 광자 한 개가 생성되고, 광자가 한 개인 상태에 $\hat{a}^\dagger$를 작용하면 광자 두 개인 상태가 되고, 광자가 두 개인 상태에 $\hat{a}^\dagger$를 작용하면 광자 세 개인 상태가 된다. 다시 말해

$$\hat{a}^\dagger |0\rangle \longrightarrow |1\rangle$$

$$\hat{a}^\dagger |1\rangle \longrightarrow |2\rangle$$

$$\hat{a}^\dagger |2\rangle \longrightarrow |3\rangle$$

과 같다. 일반적으로 광자가 n개인 상태에 $\hat{a}^\dagger$를 작용하면 광자가 $(n+1)$개인 상태가 되므로

$$\hat{a}^\dagger |n\rangle \longrightarrow |n+1\rangle$$

이다. 이제 다음과 같이 비례상수를 도입하자.

$$\hat{a}^\dagger | n \rangle = c_{n+1} | n+1 \rangle \qquad (4\text{-}1\text{-}4)$$

따라서

$$\hat{a}^\dagger | 0 \rangle = c_1 | 1 \rangle \qquad (4\text{-}1\text{-}5)$$

$$\hat{a}^\dagger | 1 \rangle = c_2 | 2 \rangle \qquad (4\text{-}1\text{-}6)$$

$$\hat{a}^\dagger | 2 \rangle = c_3 | 3 \rangle \qquad (4\text{-}1\text{-}7)$$

과 같다. 여기서 $c_1, c_2, c_3, \cdots$ 은 실수이다.

이번에는 광자수가 줄어드는 경우를 생각해 보자.

광자가 한 개인 상태에 $\hat{a}$를 작용하면 광자 0개인 상태(진공 상태)가 되고, 광자가 두 개인 상태에 $\hat{a}$를 작용하면 광자 한 개인 상태가 되고, 광자가 세 개인 상태에 $\hat{a}$를 작용하면 광자 두 개인 상태가 된다. 그러니까

$$\hat{a} | 1 \rangle \longrightarrow | 0 \rangle$$

$$\hat{a} | 2 \rangle \longrightarrow | 1 \rangle$$

$$\hat{a} | 3 \rangle \longrightarrow | 2 \rangle$$

와 같다. 일반적으로 광자가 n개인 상태에 $\hat{a}$를 작용하면 광자가 $(n-1)$개인 상태가 되므로

$$\hat{a}|n\rangle \longrightarrow |n-1\rangle$$

이다. 이제 다음과 같이 비례상수를 도입하자.

$$\hat{a}|n\rangle = d_n|n-1\rangle \qquad (4\text{-}1\text{-}8)$$

따라서

$$\hat{a}|1\rangle = d_1|0\rangle \qquad (4\text{-}1\text{-}9)$$

$$\hat{a}|2\rangle = d_2|1\rangle \qquad (4\text{-}1\text{-}10)$$

$$\hat{a}|3\rangle = d_3|2\rangle \qquad (4\text{-}1\text{-}11)$$

와 같다. 여기서 $d_1, d_2, d_3, \cdots$은 실수이다.

물리군 $c_1, c_2, c_3, \cdots$과 $d_1, d_2, d_3, \cdots$은 어떻게 결정하죠?

정교수 우선 식 (4-1-5)를 볼까?

$$\hat{a}^\dagger|0\rangle = c_1|1\rangle$$

양변의 왼쪽에 $\langle 1|$을 작용하면

$$\langle 1|\hat{a}^\dagger|0\rangle = c_1$$

이 돼. 이제 식 (4-1-9)를 봐.

$$\hat{a}\,|\,1\,\rangle = d_1\,|\,0\,\rangle$$

양변의 왼쪽에 ⟨ 0 |을 작용하면

$$\langle\,0\,|\,\hat{a}\,|\,1\,\rangle = d_1$$

이 돼. 그런데 d_1은 실수이므로 자신과 자신의 수반이 같아. 그러므로

$$d_1 = (\langle\,0\,|\cdot\hat{a}\cdot|\,1\,\rangle)^\dagger$$

$$= (|\,1\,\rangle)^\dagger \cdot (\hat{a})^\dagger \cdot (\langle\,0\,|)^\dagger$$

$$= \langle\,1\,|\cdot(\hat{a})^\dagger\cdot|\,0\,\rangle$$

$$= \langle\,1\,|\,\hat{a}^\dagger\,|\,0\,\rangle$$

$$= c_1$$

이야. 일반적으로

$$d_n = c_n \qquad\qquad (4\text{-}1\text{-}12)$$

이 되지. 따라서 다음과 같이 쓸 수 있어.

$$\hat{a}\,|\,n\,\rangle = d_n\,|\,n-1\,\rangle \qquad\qquad (4\text{-}1\text{-}13)$$

$$\hat{a}^\dagger\,|\,n\,\rangle = d_{n+1}\,|\,n+1\,\rangle \qquad\qquad (4\text{-}1\text{-}14)$$

물리군 진공 상태에 $\hat{a}$를 작용하면 어떻게 되나요?

정교수 그런 상태는 존재하지 않아. 그러니까

$$\hat{a}|0\rangle = 0$$

으로 두어야 해.

물리군 그렇다면 $d_0 = 0$이군요.

정교수 맞아.

물리군 이제 d_n을 구하는 일만 남았네요.

정교수 그렇지. 식 (4-1-2)에 $|n\rangle$을 작용하면

$$\hat{a}\hat{a}^\dagger|n\rangle - \hat{a}^\dagger\hat{a}|n\rangle = I|n\rangle \qquad (4\text{-}1\text{-}15)$$

이고,

$$\begin{aligned}
\hat{a}\hat{a}^\dagger|n\rangle &= \hat{a}(\hat{a}^\dagger|n\rangle) \\
&= \hat{a}(d_{n+1}|n+1\rangle) \\
&= d_{n+1}(\hat{a}|n+1\rangle) \\
&= d_{n+1}(d_{n+1}|n\rangle) \\
&= d_{n+1}^2|n\rangle
\end{aligned}$$

이 돼. 마찬가지로

$$\hat{a}^\dagger \hat{a} \mid n \rangle = d_n^2 \mid n \rangle$$

이지. 그러니까 식 (4-1-15)는

$$d_{n+1}^2 - d_n^2 = 1$$

이라고 쓸 수 있어. 이 식은

$$d_n^2 = n$$

이면 만족하지. 즉,

$$d_n = \sqrt{n}$$

이야. 따라서 식 (4-1-13)과 (4-1-14)는 다음과 같아.

$$\hat{a} \mid n \rangle = \sqrt{n} \mid n-1 \rangle \qquad (4\text{-}1\text{-}16)$$

$$\hat{a}^\dagger \mid n \rangle = \sqrt{n+1} \mid n+1 \rangle \qquad (4\text{-}1\text{-}17)$$

여기서 $\hat{N} = \hat{a}^\dagger \hat{a}$라고 하면

$$\hat{N} \mid n \rangle = n \mid n \rangle \qquad (4\text{-}1\text{-}18)$$

이야. 그러니까 광자가 n개인 상태를 나타내는 상태 벡터에 $\hat{N}$을 작용하면, 광자수와 광자가 n개인 상태를 나타내는 상태 벡터의 곱이 되지. 그래서 $\hat{N}$을 광자수 연산자라고 불러. 즉, 광자수 연산자의 고

윷값이 광자수라네.

벡터 대수학 _ 전기장과 자기장을 연산자로 바꾸기 위해

정교수 그럼 전자기파를 어떻게 양자화하는지 알아볼게.

물리군 양자화는 고전적인 양을 연산자로 바꾸는 과정이죠?

정교수 맞아. 그러니까 전기장과 자기장을 어떻게 연산자로 바꾸는지 살펴볼 거야. 그러기 위해서는 먼저 벡터에 대해 조금 알아야 해.

빈 공간에 전하가 있으면 주변에 전기장이라는 공간을 만든다. 어떤 위치에서 전기장의 세기를 $\vec{E}$라고 하는데, 전기장의 세기를 줄여서 전기장이라고도 한다. 우리가 3차원 공간을 생각하면 전기장은 다음과 같이 나타낼 수 있다.

$$\vec{E} = E_x \hat{i} + E_y \hat{j} + E_z \hat{k}$$

여기서 $\hat{i}, \hat{j}, \hat{k}$는 x축, y축, z축에서의 단위벡터이다.

마찬가지로 빈 공간에 전류가 있으면 주변에 자기장이라는 공간을 만든다. 어떤 위치에서 자기장의 세기를 $\vec{B}$라고 하는데, 자기장의 세기를 줄여서 자기장이라고도 한다. 우리가 3차원 공간을 생각하면 자기장은 다음과 같이 나타낼 수 있다.

$$\vec{B} = B_x\hat{i} + B_y\hat{j} + B_z\hat{k}$$

일반적으로 전기장과 자기장은 공간과 시간의 함수이다. 즉,

$$\vec{E} = \vec{E}(x, y, z, t)$$

$$\vec{B} = \vec{B}(x, y, z, t)$$

이다.

이제 몇 가지 기호를 복습하겠다. 두 벡터

$$\vec{A} = A_x\hat{i} + A_y\hat{j} + A_z\hat{k}$$

$$\vec{B} = B_x\hat{i} + B_y\hat{j} + B_z\hat{k}$$

를 생각하자. 이때 두 벡터의 내적을

$$\vec{A} \cdot \vec{B} = A_xB_x + A_yB_y + A_zB_z$$

로 정의한다. 만일 두 벡터의 내적이 0이면 두 벡터는 수직으로 만난다.

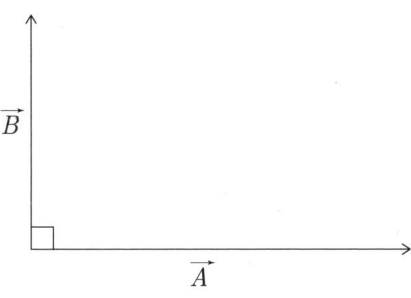

이번에는 두 벡터의 외적을 알아보자. 두 벡터의 외적은

$$\vec{A} \times \vec{B}$$

로 나타내는데 그 크기는

$$|\vec{A} \times \vec{B}| = |\vec{A}||\vec{B}|\sin\theta$$

이고, 방향은 $\vec{A}$에서 $\vec{B}$로 오른손을 감아쥐었을 때 엄지손가락이 가리키는 쪽이다.

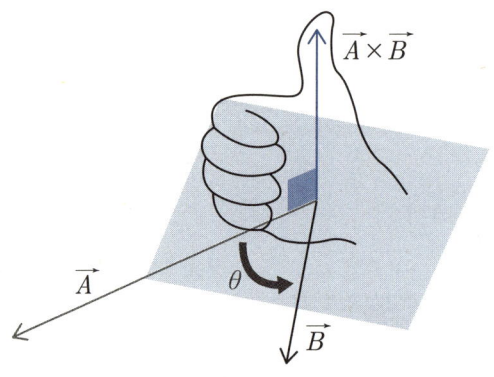

여기서 θ는 두 벡터가 이루는 사잇각이다. 그러므로 같은 두 벡터의 외적은 0이다.

$$\vec{A} \times \vec{A} = 0$$

두 벡터의 외적을

$$\vec{A} \times \vec{B} = (\vec{A} \times \vec{B})_x \hat{i} + (\vec{A} \times \vec{B})_y \hat{j} + (\vec{A} \times \vec{B})_z \hat{k}$$

라고 하면

$$(\vec{A} \times \vec{B})_x = A_y B_z - A_z B_y$$

$$(\vec{A} \times \vec{B})_y = A_z B_x - A_x B_z$$

$$(\vec{A} \times \vec{B})_z = A_x B_y - A_y B_x$$

가 된다.

이제 벡터 미분 연산자에 대해 알아보자. 벡터 미분 연산자는

$$\vec{\nabla} = \hat{i} \frac{\partial}{\partial x} + \hat{j} \frac{\partial}{\partial y} + \hat{k} \frac{\partial}{\partial z}$$

로 정의한다. 이것을 좀 더 간단히

$$\vec{\nabla} = \hat{i} \partial_x + \hat{j} \partial_y + \hat{k} \partial_z$$

라고 쓰기도 한다. 여기서 ∂_x는 x에 대한 편미분, ∂_y는 y에 대한 편미분, ∂_z는 z에 대한 편미분을 나타낸다. 편미분은 그 변수만 문자로 생각하는 미분이다. 예를 들어

$$\frac{d}{dx}(2x) = 2$$

이듯이

$$\partial_x(xyz) = \partial_x[(yz)x] = yz$$

$$\partial_y(xyz) = \partial_y[(zx)y] = zx$$

$$\partial_z(xyz) = \partial_z[(xy)z] = xy$$

로 계산한다.

벡터 미분 연산자 $\vec{\nabla}$와 임의의 벡터 $\vec{A}$에 대한 내적과 외적은

$$\vec{\nabla} \cdot \vec{A} = \frac{\partial A_x}{\partial x} + \frac{\partial A_y}{\partial y} + \frac{\partial A_z}{\partial z} = \partial_x A_x + \partial_y A_y + \partial_z A_z$$

$$\vec{\nabla} \times \vec{A} = (\vec{\nabla} \times \vec{A})_x \hat{i} + (\vec{\nabla} \times \vec{A})_y \hat{j} + (\vec{\nabla} \times \vec{A})_z \hat{k}$$

이고, 이때

$$(\vec{\nabla} \times \vec{A})_x = \partial_y A_z - \partial_z A_y$$

$$(\vec{\nabla} \times \vec{A})_y = \partial_z A_x - \partial_x A_z$$

$$(\vec{\nabla} \times \vec{A})_z = \partial_x A_y - \partial_y A_x$$

이다.

맥스웰 방정식 _ 전기장과 자기장을 묘사하는 복잡한 방정식

정교수 전기장과 자기장을 만족하는 방정식을 맥스웰 방정식이라고 불러. 전하도 없고 전류도 없는 공간에서 맥스웰 방정식은 다음과 같아.

$$\vec{\nabla} \cdot \vec{E} = 0 \qquad (4\text{-}3\text{-}1)$$

$$\vec{\nabla} \times \vec{E} = -\frac{\partial \vec{B}}{\partial t} \qquad (4\text{-}3\text{-}2)$$

$$\vec{\nabla} \cdot \vec{B} = 0 \qquad (4\text{-}3\text{-}3)$$

$$\vec{\nabla} \times \vec{B} = \frac{1}{c^2}\frac{\partial \vec{E}}{\partial t} \qquad (4\text{-}3\text{-}4)$$

물리군 굉장히 복잡한 방정식이네요.

정교수 그렇지. 네 개의 방정식 중에서 식 (4-3-3)을 봐. 여기서

$$\vec{B} = \vec{\nabla} \times \vec{A} \qquad (4\text{-}3\text{-}5)$$

라고 하면 식 (4-3-3)을 만족해.

$$\vec{\nabla} \cdot \vec{B} = \vec{\nabla} \cdot \vec{\nabla} \times \vec{A} = \vec{\nabla} \times \vec{\nabla} \cdot \vec{A} = 0$$

이때 $\vec{A}$를 벡터퍼텐셜이라고 불러.

물리군 $\vec{\nabla} \times \vec{\nabla} = 0$이기 때문에 식이 성립하는군요.

정교수 맞아.

물리군 자기장은 벡터퍼텐셜로 나타낼 수 있네요.

정교수 $\vec{B} = \vec{\nabla} \times \vec{A}$에서 $\vec{A}$에 $\vec{\nabla} f(x, y, z, t)$를 더해도 자기장은 달라지지 않아. 여기서 $f(x, y, z, t)$는 임의의 함수야.

물리군 마찬가지로 $\vec{\nabla} \times \vec{\nabla} = 0$이기 때문이군요.

정교수 $f(x, y, z, t)$를 잘 선택해서 $\vec{A}$가 어떤 관계식을 만족하도록 할 수 있어. 물리학자들이 주로 사용하는 관계식은

$$\vec{\nabla} \cdot \vec{A} = 0 \qquad (4\text{-}3\text{-}6)$$

이지. 이 조건을 만족하도록 $f(x, y, z, t)$를 선택하면 전기장도 벡터퍼텐셜로만 나타낼 수 있어.

$$\vec{E} = -\frac{\partial}{\partial t} \vec{A} \qquad (4\text{-}3\text{-}7)$$

이때

$$\vec{\nabla} \cdot \vec{E} = -\frac{\partial}{\partial t}(\vec{\nabla} \cdot \vec{A}) = 0$$

이 되어 식 (4-3-1)이 성립하고,

$$\vec{\nabla} \times \vec{E} = -\frac{\partial}{\partial t}(\vec{\nabla} \times \vec{A}) = -\frac{\partial}{\partial t}\vec{B}$$

가 되어 식 (4-3-2)가 성립하지.

물리군 전기장과 자기장 모두 벡터퍼텐셜만으로 나타낼 수 있군요.

정교수 물론이야. 식 (4-3-5)와 (4-3-7)을 식 (4-3-4)에 넣으면

$$\vec{\nabla} \times (\vec{\nabla} \times \vec{A}) = -\frac{1}{c^2}\frac{\partial^2 \vec{A}}{\partial t^2}$$

이 돼. 이제 $\vec{\nabla} \times (\vec{\nabla} \times \vec{A})$의 x성분을 구해볼게.

$$\left[\vec{\nabla} \times (\vec{\nabla} \times \vec{A})\right]_x = \frac{\partial (\vec{\nabla} \times \vec{A})_z}{\partial y} - \frac{\partial (\vec{\nabla} \times \vec{A})_y}{\partial z}$$

$$= \frac{\partial}{\partial y}\left(\frac{\partial A_y}{\partial x} - \frac{\partial A_x}{\partial y}\right) - \frac{\partial}{\partial z}\left(\frac{\partial A_x}{\partial z} - \frac{\partial A_z}{\partial x}\right)$$

$$= \frac{\partial}{\partial x}\left(\frac{\partial A_y}{\partial y} + \frac{\partial A_z}{\partial z}\right) - \frac{\partial^2 A_x}{\partial y^2} - \frac{\partial^2 A_x}{\partial z^2}$$

그런데

$$\vec{\nabla} \cdot \vec{A} = \frac{\partial A_x}{\partial x} + \frac{\partial A_y}{\partial y} + \frac{\partial A_z}{\partial z} = 0$$

이므로

$$\left[\vec{\nabla} \times (\vec{\nabla} \times \vec{A})\right]_x = \frac{\partial}{\partial x}\left(-\frac{\partial A_x}{\partial x}\right) - \frac{\partial^2 A_x}{\partial y^2} - \frac{\partial^2 A_x}{\partial z^2}$$

$$= -\frac{\partial^2 A_x}{\partial x^2} - \frac{\partial^2 A_x}{\partial y^2} - \frac{\partial^2 A_x}{\partial z^2}$$

이야. 그러니까 벡터퍼텐셜의 x성분은

$$\frac{\partial^2 A_x}{\partial x^2}+\frac{\partial^2 A_x}{\partial y^2}+\frac{\partial^2 A_x}{\partial z^2}-\frac{1}{c^2}\frac{\partial^2 A_x}{\partial t^2}=0$$

을 만족하고, 마찬가지로 벡터퍼텐셜의 y, z성분도

$$\frac{\partial^2 A_y}{\partial x^2}+\frac{\partial^2 A_y}{\partial y^2}+\frac{\partial^2 A_y}{\partial z^2}-\frac{1}{c^2}\frac{\partial^2 A_y}{\partial t^2}=0$$

$$\frac{\partial^2 A_z}{\partial x^2}+\frac{\partial^2 A_z}{\partial y^2}+\frac{\partial^2 A_z}{\partial z^2}-\frac{1}{c^2}\frac{\partial^2 A_z}{\partial t^2}=0$$

을 만족해. 이것을 벡터 형태로 쓰면

$$\frac{\partial^2 \vec{A}}{\partial x^2}+\frac{\partial^2 \vec{A}}{\partial y^2}+\frac{\partial^2 \vec{A}}{\partial z^2}-\frac{1}{c^2}\frac{\partial^2 \vec{A}}{\partial t^2}=0 \qquad (4\text{-}3\text{-}8)$$

이 되지.

파동방정식과 전자기파 _ 파동은 어떻게 나타낼까?

정교수 식 (4-3-8)은 파동방정식으로 잘 알려져 있어. 일반적으로 전파속도가 c인 파동방정식은 다음과 같아.

$$\frac{\partial^2 u}{\partial x^2} + \frac{\partial^2 u}{\partial y^2} + \frac{\partial^2 u}{\partial z^2} - \frac{1}{c^2}\frac{\partial^2 u}{\partial t^2} = 0 \qquad (4\text{-}4\text{-}1)$$

이것은 3차원 공간의 파동방정식이야. 조금 쉽게 이해하기 위해 1차원 공간에서 파동방정식을 살펴볼게.

$$\frac{\partial^2 u}{\partial x^2} - \frac{1}{c^2}\frac{\partial^2 u}{\partial t^2} = 0 \qquad (4\text{-}4\text{-}2)$$

이때 u는 x와 t의 함수야. 즉, $u(x, t)$이지.
파동의 일반적인 모습은 사인함수 모양이야. 식으로 나타내면

$$u(x, t) = \sin(kx - wt)$$

의 꼴이라네. 여기서 x에 대한 주기를 파동의 파장이라 하고 λ로 표현해. 그러니까

$$u(x + \lambda, t) = u(x, t)$$

이지. 즉,

$$\sin(k(x + \lambda) - wt) = \sin(kx - wt)$$

가 되어,

$$k\lambda = 2\pi$$

또는

$$k = \frac{2\pi}{\lambda}$$

임을 알 수 있어. k를 파동의 파수라고 부르지.

마찬가지로 시간 t에 대한 주기를 파동의 주기라 하고 T로 쓴다네. 그러니까

$$u(x, t+T) = u(x, t)$$

이지. 그러므로

$$wT = 2\pi$$

또는

$$w = \frac{2\pi}{T}$$

임을 알 수 있어. 이때 w를 각진동수라고 불러.

파동의 파장과 주기와 전파속도 사이에는 다음 관계가 성립해.

$$\lambda = cT$$

다시 말해

$$w = ck$$

가 성립하지.

한편 $u(x, t) = \sin(kx - wt)$는

$$\frac{\partial^2 u}{\partial x^2} = -k^2 u$$

$$\frac{\partial^2 u}{\partial t^2} = -w^2 u$$

를 만족하고, $w = ck$이니까 $u(x, t) = \sin(kx - wt)$는 파동방정식

$$\frac{\partial^2 u}{\partial x^2} - \frac{1}{c^2}\frac{\partial^2 u}{\partial t^2} = 0$$

의 해가 돼. 그런데 사인함수가 아니라 코사인함수로 써도 파동방정식을 만족해. 즉,

$$u = \cos(kx - wt)$$

도 파동방정식의 해가 되는 거야. 따라서 파동방정식 (4-4-2)의 일반해는

$$u = c_1\cos(kx - wt) + c_2\sin(kx - wt) \qquad (4\text{-}4\text{-}3)$$

로 나타낼 수 있어. 여기서 c_1, c_2는 임의의 실수야.

식 (4-4-3)을 간단히 파동이라고도 부르는데, 이걸 복소수로 바꿔볼게.

물리군 실수인 파동을 어떻게 복소수로 바꾸나요?

정교수 차근차근 생각해 볼까? e^{Ax}을 미분하면 어떻게 되지?

물리군 Ae^{Ax}이에요.

정교수 맞아. 이 성질을 이용하면

$$(e^{ikx})' = ike^{ikx}$$

이야. 여기서 $e^{ikx} = f(x) + ig(x)$라고 하면

$$f' + ig' = ik(f + ig)$$

가 되고, 실수부와 허수부를 비교하면

$$f' = -kg$$

$$g' = kf$$

인 걸 알 수 있어. 이 식을 만족하는 두 함수는

$$f = \cos kx$$

$$g = \sin kx$$

이므로 다음과 같아.

$$e^{ikx} = \cos kx + i \sin kx$$

여기에 k 대신 $-k$를 넣으면

$$e^{-ikx} = \cos kx - i \sin kx$$

가 돼. 두 식을 더하면

$$\cos kx = \frac{1}{2}(e^{ikx} + e^{-ikx})$$

이고, 두 식을 빼면

$$\sin kx = \frac{1}{2i}(e^{ikx} - e^{-ikx})$$

이야. 이 공식들을 이용하면 파동 (4-4-3)은

$$u = c_1 \left[\frac{1}{2}(e^{i(kx-wt)} + e^{-i(kx-wt)}) \right] + c_2 \left[\frac{1}{2i}(e^{i(kx-wt)} - e^{-i(kx-wt)}) \right]$$

$$= \left(\frac{c_1}{2} + \frac{c_2}{2i} \right) e^{i(kx-wt)} + \left(\frac{c_1}{2} - \frac{c_2}{2i} \right) e^{-i(kx-wt)}$$

가 돼. 이 식에서

$$A = \frac{c_1}{2} + \frac{c_2}{2i}$$

라고 두면, 이것의 켤레복소수는

$$A^* = \frac{c_1}{2} - \frac{c_2}{2i}$$

이니까

$$u = Ae^{i(kx-wt)} + A^*e^{-i(kx-wt)} \tag{4-4-4}$$

으로 나타낼 수 있어.

마찬가지로 3차원에서의 파동은

$$u = Ae^{i(k_xx+k_yy+k_zz-wt)} + A^*e^{-i(k_xx+k_yy+k_zz-wt)} \tag{4-4-5}$$

이지. 이때 파수벡터를

$$\vec{k} = k_x\hat{i} + k_y\hat{j} + k_z\hat{k}$$

라 하고, 위치벡터를

$$\vec{r} = x\hat{i} + y\hat{j} + z\hat{k}$$

라고 하면, 식 (4-4-5)는 다음과 같이 쓸 수 있어.

$$u = Ae^{i(\vec{k}\cdot\vec{r}-wt)} + A^*e^{-i(\vec{k}\cdot\vec{r}-wt)}$$

여기서 파수벡터의 크기를 k라고 하는데

$$k = |\vec{k}| = \sqrt{k_x^2 + k_y^2 + k_z^2}$$

이고, 파장 λ는

$$\lambda = \frac{2\pi}{k}$$

가 되지. 이때 파동의 진행 방향은 $\vec{k}$의 방향이야. 그리고

$$w = ck$$

이지.

전자기파 _ 벡터퍼텐셜이 만족하는 파동방정식에서

정교수 이제 벡터퍼텐셜이 만족하는 파동방정식 (4-3-8)을 봐. 이 경우 $\vec{A}$는 벡터이고 파동이므로

$$\vec{A}(\vec{r}, t) = \vec{A_0} e^{i(\vec{k}\cdot\vec{r}-wt)} + \vec{A_0}^* e^{-i(\vec{k}\cdot\vec{r}-wt)} \tag{4-5-1}$$

이 돼. 여기서 $\vec{A_0}$은 복소수 상수 벡터이고, $\vec{A_0}^*$은 $\vec{A_0}$의 켤레야. 즉,

$$\vec{A_0} = A_{0x}\hat{i} + A_{0y}\hat{j} + A_{0z}\hat{k}$$

이고 A_{0x}, A_{0y}, A_{0z}는 복소수라네. 그리고

$$\vec{A_0}^* = A_{0x}^*\hat{i} + A_{0y}^*\hat{j} + A_{0z}^*\hat{k}$$

이지.

식 (4-5-1)은 $\vec{\nabla} \cdot \vec{A} = 0$을 만족해야 해. 여기서

$$\vec{\nabla} \cdot \left[\vec{A_0} e^{i(\vec{k} \cdot \vec{r} - wt)} \right]$$

$$= \partial_x \left(A_{0x} e^{i(k_x x + k_y y + k_z z - wt)} \right) + \partial_y \left(A_{0y} e^{i(k_x x + k_y y + k_z z - wt)} \right) + \partial_z \left(A_{0z} e^{i(k_x x + k_y y + k_z z - wt)} \right)$$

$$= ik_x A_{0x} e^{i(k_x x + k_y y + k_z z - wt)} + ik_y A_{0y} e^{i(k_x x + k_y y + k_z z - wt)} + ik_z A_{0z} e^{i(k_x x + k_y y + k_z z - wt)}$$

$$= i(k_x A_{0x} + k_y A_{0y} + k_z A_{0z}) e^{i(\vec{k} \cdot \vec{r} - wt)}$$

$$= i(\vec{k} \cdot \vec{A_0}) e^{i(\vec{k} \cdot \vec{r} - wt)}$$

이고

$$\vec{\nabla} \cdot \left[\vec{A_0}^* e^{-i(\vec{k} \cdot \vec{r} - wt)} \right] = -i(\vec{k} \cdot \vec{A_0}^*) e^{-i(\vec{k} \cdot \vec{r} - wt)}$$

이므로

$$\vec{\nabla} \cdot \vec{A} = i(\vec{k} \cdot \vec{A_0}) e^{i(\vec{k} \cdot \vec{r} - wt)} - i(\vec{k} \cdot \vec{A_0}^*) e^{-i(\vec{k} \cdot \vec{r} - wt)} = 0$$

이 돼. 따라서 다음과 같아.

$$\vec{k} \cdot \vec{A_0} = 0 \qquad\qquad (4\text{-}5\text{-}2)$$

$$\vec{k} \cdot \vec{A_0}^* = 0 \qquad\qquad (4\text{-}5\text{-}3)$$

물리군 파수벡터는 $\vec{A_0}$과 $\vec{A_0}^*$에 수직이군요.

정교수 맞아. 이제 전기장과 자기장을 벡터퍼텐셜로 나타내면

$$\vec{E} = iw\left[\vec{A_0}e^{i(\vec{k}\cdot\vec{r}-wt)} - \vec{A_0}^*e^{-i(\vec{k}\cdot\vec{r}-wt)}\right] \tag{4-5-4}$$

$$\vec{B} = i\vec{k}\times\left[\vec{A_0}e^{i(\vec{k}\cdot\vec{r}-wt)} - \vec{A_0}^*e^{-i(\vec{k}\cdot\vec{r}-wt)}\right] \tag{4-5-5}$$

이야. 식 (4-5-2)와 (4-5-3)을 이용하면

$$\vec{k}\cdot\vec{E} = 0 \tag{4-5-6}$$

$$\vec{k}\cdot\vec{B} = 0 \tag{4-5-7}$$

임을 알 수 있어.

물리군 파수벡터는 전기장과 자기장에 수직이네요.

정교수 그렇지. 이때

$$\vec{E}\cdot\vec{B} = -w\left[\vec{A_0}\cdot(\vec{k}\times\vec{A_0})e^{2i(\vec{k}\cdot\vec{r}-wt)} - \vec{A_0}^*\cdot(\vec{k}\times\vec{A_0}) - \vec{A_0}\cdot(\vec{k}\times\vec{A_0}^*)\right.$$
$$\left. + \vec{A_0}^*\cdot(\vec{k}\times\vec{A_0}^*)e^{-2i(\vec{k}\cdot\vec{r}-wt)}\right]$$

이고, 여기서 $\vec{k} = k\hat{k}$ 라고 하면

$$(\vec{k}\times\vec{A_0}) = k(A_{0x}\hat{j} - A_{0x}\hat{i})$$

$$(\vec{k}\times\vec{A_0}^*) = k(A_{0x}^*\hat{j} - A_{0x}^*\hat{i})$$

이지. 따라서

$$\vec{A_0} \cdot (\vec{k} \times \vec{A_0}) = k[A_{0x}(-A_{0y}) + A_{0y}A_{0x}] = 0$$

이고 마찬가지로

$$\vec{A_0}^* \cdot (\vec{k} \times \vec{A_0}^*) = 0$$

이 돼. 또한

$$\vec{A_0}^* \cdot (\vec{k} \times \vec{A_0}) = k(-A_{0x}^* A_{0y} + A_{0y}^* A_{0x})$$

$$\vec{A_0} \cdot (\vec{k} \times \vec{A_0}^*) = k(-A_{0x}A_{0y}^* + A_{0y}A_{0x}^*)$$

이므로

$$\vec{A_0}^* \cdot (\vec{k} \times \vec{A_0}) + \vec{A_0} \cdot (\vec{k} \times \vec{A_0}^*) = 0$$

이 되어,

$$\vec{E} \cdot \vec{B} = 0$$

인 것을 알 수 있어. 즉, $\vec{k}, \vec{E}, \vec{B}$는 서로 수직이라는 거지. 이때 $\vec{k}$는 파동의 전파 방향이야.

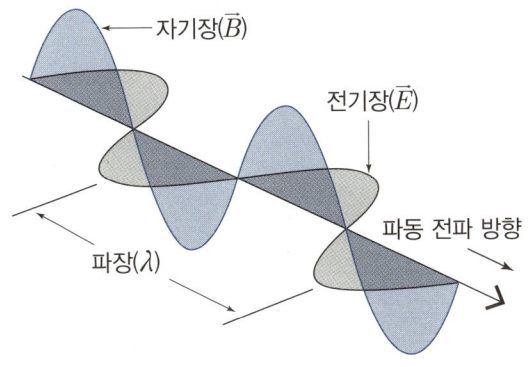

전자기장의 양자화 _ 입체 속의 전자기파

정교수 그럼 부피가 V인 입체 속의 전자기파를 생각해 볼까?

전자기파의 에너지를 H라고 하면

$$H = \frac{V}{2}\left(\epsilon_0 \langle |\vec{E}|^2 \rangle_T + \frac{1}{\mu_0} \langle |\vec{B}|^2 \rangle_T \right) \tag{4-6-1}$$

이다. 여기서

$$\frac{1}{\epsilon_0 \mu_0} = c^2$$

이고, $\langle\ \rangle_T$는 주기 T 동안의 평균을 의미한다. 먼저 $\langle |\vec{E}|^2 \rangle_T$를 계산하자.

$$|\vec{E}|^2 = \vec{E} \cdot \vec{E}^*$$

$$= iw\left[\vec{A_0}e^{i(\vec{k}\cdot\vec{r}-wt)} - \vec{A_0}^*e^{-i(\vec{k}\cdot\vec{r}-wt)}\right]\cdot(-iw)\left[\vec{A_0}^*e^{-i(\vec{k}\cdot\vec{r}-wt)} - \vec{A_0}e^{i(\vec{k}\cdot\vec{r}-wt)}\right]$$

$$= 2w^2|\vec{A_0}|^2 + w^2(\vec{A_0}\cdot\vec{A_0}e^{-2iwt}e^{2i\vec{k}\cdot\vec{r}} + \vec{A_0}^*\cdot\vec{A_0}^*e^{2iwt}e^{-2i\vec{k}\cdot\vec{r}})$$

이고,

$$\langle e^{-2iwt}\rangle_T = \frac{1}{T}\int_0^T e^{-2iwt}dt$$

$$= \frac{1}{-2iw}(e^{-2iwT} - 1)$$

이다. 이때

$$w = \frac{2\pi}{T}$$

이므로

$$e^{-2iwT} = e^{-4\pi i} = \cos 4\pi - i\sin 4\pi = 1$$

이 되어,

$$\langle e^{-2iwt}\rangle_T = 0$$

이다. 마찬가지로

$$\langle e^{2iwt} \rangle_T = 0$$

임을 알 수 있다. 그러므로

$$\langle |\vec{E}|^2 \rangle_T = \frac{1}{T}\int_0^T 2w^2 |\vec{A_0}|^2 \, dt$$

$$= \frac{1}{T} 2w^2 |\vec{A_0}|^2 T$$

$$= 2w^2 |\vec{A_0}|^2 \tag{4-6-2}$$

이 된다. 같은 방법으로

$$\langle |\vec{B}|^2 \rangle_T = 2k^2 |\vec{A_0}|^2 \tag{4-6-3}$$

이다. 따라서 전자기파의 에너지는 다음과 같다.

$$H = V\left(\epsilon_0 w^2 |\vec{A_0}|^2 + \frac{k^2}{\mu_0} |\vec{A_0}|^2\right)$$

한편

$$\frac{k^2}{\mu_0} = \epsilon_0 \times \frac{1}{\epsilon_0 \mu_0} k^2 = \epsilon_0 \times c^2 k^2 = \epsilon_0 w^2$$

이므로

$$H = 2V\epsilon_0 w^2 |\vec{A_0}|^2 \tag{4-6-4}$$

이 된다. 이제 $\vec{A_0}$방향의 단위벡터를 $\hat{e}$라 하고

$$\vec{A_0} = \frac{\hat{e}}{\sqrt{4\epsilon_0 w^2 V}}(wx + ip)$$

로 두자. 이때

$$\vec{A_0}^* = \frac{\hat{e}}{\sqrt{4\epsilon_0 w^2 V}}(wx - ip)$$

로부터 에너지는

$$H = \frac{1}{2}p^2 + \frac{1}{2}w^2 x^2 \tag{4-6-5}$$

이 된다. 이것은 질량이 1이고 퍼텐셜에너지가 $\frac{1}{2}w^2 x^2$인 입자의 역학적에너지이다. 그러므로 x와 p를 연산자 $\hat{x}, \hat{p}$로 바꾸어

$$[\hat{x}, \hat{p}] = i\hbar$$

가 되게 하려면 역학적에너지는 에너지 연산자 $\hat{H}$가 되고

$$\hat{H} = \frac{1}{2}\hat{p}^2 + \frac{1}{2}w^2 \hat{x}^2 \tag{4-6-6}$$

이다. 여기서 다음과 같이 연산자를 도입하자.

$$\hat{a} = \frac{1}{\sqrt{2\hbar w}}(w\hat{x} + i\hat{p})$$

$$\hat{a}^\dagger = \frac{1}{\sqrt{2\hbar w}}(w\hat{x} - i\hat{p}) \tag{4-6-7}$$

이 경우

$$[\hat{a}, \hat{a}^\dagger] = 1 \tag{4-6-8}$$

이고, 에너지 연산자는

$$\hat{H} = \hbar w \left(\hat{a}^\dagger \hat{a} + \frac{1}{2}\right) \tag{4-6-9}$$

이다.

이제 광자수 연산자 $\hat{N} = \hat{a}^\dagger \hat{a}$에 대해 광자가 n개인 상태를 나타내는 상태 벡터 $|n\rangle$을 도입하면

$$\hat{N}|n\rangle = n|n\rangle \quad (n = 0, 1, 2, 3, \cdots)$$

이고,

$$\hat{H}|n\rangle = \hbar w \left(n + \frac{1}{2}\right)|n\rangle \quad (n = 0, 1, 2, 3, \cdots) \tag{4-6-10}$$

이 된다.

한편 소멸 연산자와 생성 연산자는

$$\hat{a}|n\rangle = \sqrt{n}\,|n-1\rangle$$

$$\hat{a}^\dagger|n\rangle = \sqrt{n+1}\,|n+1\rangle \qquad (4\text{-}6\text{-}11)$$

을 만족한다. 이때 진공 상태(광자가 한 개도 없는 상태)는 $|0\rangle$으로 묘사하고

$$a|0\rangle = 0$$

에 의해 정의된다.

이제 벡터퍼텐셜, 전기장, 자기장은 연산자로 바뀐다. 즉, $\vec{A}, \vec{E}, \vec{B}$에 대한 연산자를 $\hat{A}, \hat{E}, \hat{B}$라고 하면 다음과 같다.

$$\hat{A} = \hat{e}\sqrt{\frac{\hbar}{2\epsilon_0 w V}}\,(\hat{a}e^{i(\vec{k}\cdot\vec{r}-wt)} + \hat{a}^\dagger e^{-i(\vec{k}\cdot\vec{r}-wt)})$$

$$\hat{E} = i\hat{e}\sqrt{\frac{\hbar w}{2\epsilon_0 V}}\,(\hat{a}e^{i(\vec{k}\cdot\vec{r}-wt)} - \hat{a}^\dagger e^{-i(\vec{k}\cdot\vec{r}-wt)})$$

$$\hat{B} = i\vec{k}\times\hat{e}\sqrt{\frac{\hbar}{2\epsilon_0 w V}}\,(\hat{a}e^{i(\vec{k}\cdot\vec{r}-wt)} - \hat{a}^\dagger e^{-i(\vec{k}\cdot\vec{r}-wt)})$$

다섯 번째 만남

원자와 빛의 상호작용

제인스-커밍스 모형 _ 바닥상태와 첫 번째 들뜬상태

정교수 지금부터는 빛과 물질의 상호작용을 설명하려고 해. 이 이론을 처음 제창한 과학자는 제인스와 그의 제자 커밍스(Frederick Williams Cummings, 1931~2019)야. 그래서 이 모형을 제인스-커밍스 모형이라고 불러.

제인스(Edwin Thompson Jaynes, 1922~1998)

물질은 원자로 이루어져 있고 원자 속의 전자가 가질 수 있는 에너지는 무한히 많다. 하지만 제인스와 커밍스는 원자 속 에너지 상태에 대해 바닥상태와 첫 번째 들뜬상태만을 고려하는 모형을 생각했다.

제인스-커밍스 모형을 자세히 살펴보자. 전자기장과 상호작용 하지 않을 때, 원자 속 전자의 바닥상태 에너지를 E_1, 첫 번째 들뜬상태의 에너지를 E_2라고 하자. 그리고 이 에너지에 대응하는 상태를 각각 $|1\rangle$, $|2\rangle$라고 하자. 전자기장과 상호작용 하지 않을 때 원자 속 전자

의 에너지 연산자를 H_0이라고 하면

$$H_0 \,|\, 1 \rangle = E_1 \,|\, 1 \rangle \qquad (5\text{-}1\text{-}1)$$

$$H_0 \,|\, 2 \rangle = E_2 \,|\, 2 \rangle \qquad (5\text{-}1\text{-}2)$$

가 된다.

한편 전자기장과 상호작용 할 때 전자의 총에너지 연산자는

$$H = H_0 - \frac{e}{mc}\vec{A}\cdot\vec{p} \qquad (5\text{-}1\text{-}3)$$

로 주어진다. 여기서 $\vec{A}$는 벡터퍼텐셜이다. 이 식에 대해 궁금한 독자는 이 시리즈의 《반입자》를 참고하라.

이제

$$H' = -\frac{e}{mc}\vec{A}\cdot\vec{p} \qquad (5\text{-}1\text{-}4)$$

로 놓으면

$$H = H_0 + H'$$

이다.

따라서 슈뢰딩거 방정식은

$$i\hbar\frac{\partial}{\partial t}\,|\,\psi(t)\rangle = (H_0 + H')\,|\,\psi(t)\rangle \qquad (5\text{-}1\text{-}5)$$

이고, 이때

$$|\psi(t)\rangle = a_1(t)e^{-\frac{i}{\hbar}E_1 t}|1\rangle + a_2(t)e^{-\frac{i}{\hbar}E_2 t}|2\rangle \qquad (5-1-6)$$

로 놓을 수 있다. 식 (5-1-6)을 슈뢰딩거 방정식에 넣자. 그러면 좌변은

$$i\hbar\frac{\partial}{\partial t}|\psi(t)\rangle = (i\hbar\dot{a}_1 + E_1 a_1)e^{-\frac{i}{\hbar}E_1 t}|1\rangle + (i\hbar\dot{a}_2 + E_2 a_2)e^{-\frac{i}{\hbar}E_2 t}|2\rangle$$

이고 우변은

$$(H_0 + H')|\psi(t)\rangle$$
$$= E_1 a_1 e^{-\frac{i}{\hbar}E_1 t}|1\rangle + a_1 e^{-\frac{i}{\hbar}E_1 t}H'|1\rangle + E_2 a_2 e^{-\frac{i}{\hbar}E_2 t}|2\rangle + a_2 e^{-\frac{i}{\hbar}E_2 t}H'|2\rangle$$

가 되어,

$$i\hbar\dot{a}_1 e^{-\frac{i}{\hbar}E_1 t}|1\rangle + i\hbar\dot{a}_2 e^{-\frac{i}{\hbar}E_2 t}|2\rangle = a_1 e^{-\frac{i}{\hbar}E_1 t}H'|1\rangle + a_2 e^{-\frac{i}{\hbar}E_2 t}H'|2\rangle$$
$$(5-1-7)$$

가 성립한다. 여기서 $\dot{a}_1 = \frac{da_1}{dt}, \dot{a}_2 = \frac{da_2}{dt}$이고, H'은 전자기장과 상호작용을 나타내며 원자 속 전자의 상태를 바꾸는 역할을 한다. 즉, 빛의 흡수는 바닥상태의 전자를 들뜬상태로 바꾸고, 빛의 방출은 들뜬상태의 전자를 바닥상태로 바꾼다. 그러므로 다음과 같다.

$$H'|1\rangle \longrightarrow |2\rangle$$

$$H'|2\rangle \longrightarrow |1\rangle$$

식 (5-1-7)의 왼쪽에 ⟨ 1 |을 작용하면

$$\dot{a}_1 = -\frac{i}{\hbar}\langle 1|H'|2\rangle e^{-\frac{i}{\hbar}(E_2 - E_1)t} a_2 \qquad (5\text{-}1\text{-}8)$$

이고, 식 (5-1-7)의 왼쪽에 ⟨ 2 |를 작용하면

$$\dot{a}_2 = -\frac{i}{\hbar}\langle 2|H'|1\rangle e^{\frac{i}{\hbar}(E_2 - E_1)t} a_1 \qquad (5\text{-}1\text{-}9)$$

이다. $t = 0$일 때 전자가 | 1 ⟩에 있었다면

$a_1(0) = 1$

$a_2(0) = 0$

이 된다.

이제

$$\vec{A}_0 = \hat{\epsilon} A_0$$

$$\vec{A}_0^{\,*} = \hat{\epsilon} A_0^{\,*}$$

이라고 하면

$$\vec{A}(\vec{r},t) = \hat{\epsilon}\left[A_0 e^{i(\vec{k}\cdot\vec{r}-wt)} + A_0^* e^{-i(\vec{k}\cdot\vec{r}-wt)}\right]$$

이 된다.

실제 빛의 파장은 원자 크기에 비해 아주 크다. 예를 들어 붉은빛의 경우 파장은 $\lambda_{red} \approx 650\text{nm}$로 원자 크기($\approx 0.1\text{nm}$)에 비해 엄청나게 크다. 그러므로

$$e^{i\vec{k}\cdot\vec{r}} = 1 + i\vec{k}\cdot\vec{r} + \cdots$$

$$= 1 + i|\vec{k}\|\vec{r}|\cos\phi + \cdots$$

$$= 1 + i\frac{2\pi}{\lambda_{red}}|\vec{r}|\cos\phi + \cdots$$

에서

$$\frac{1}{\lambda_{red}}|\vec{r}| \sim \frac{0.1}{650}$$

은 아주 작은 값이 되어 무시하면 다음과 같은 근사를 사용할 수 있다.

$$e^{i\vec{k}\cdot\vec{r}} \approx 1$$

$$e^{-i\vec{k}\cdot\vec{r}} \approx 1$$

따라서

$$\vec{A}(\vec{r},t) \approx \hat{\epsilon}[A_0 e^{-iwt} + A_0^* e^{iwt}]$$

으로 근사한다. 이제 다음 항을 보자.

$$\langle 1 | H' | 2 \rangle e^{-iw_0 t}$$

여기서

$$w_0 = \frac{E_2 - E_1}{\hbar}$$

로 놓자. w가 w_0과 거의 비슷한 경우만 생각할 때, 항 $e^{-i(w+w_0)t}$을 시간 한 주기에 대해 평균하면 0이 된다. 그러므로

$$I_{12} = \langle 1 | H' | 2 \rangle e^{-iw_0 t}$$

$$= -\frac{e}{mc} A_0^* e^{i(w-w_0)t} \hat{\epsilon} \cdot \langle 1 | \vec{p} | 2 \rangle$$

이다.

한편 에렌페스트 정리에 의해

$$\frac{d\vec{r}}{dt} = \frac{i}{\hbar}[H_0, \vec{r}] = \frac{\vec{p}}{m}$$

이므로

$$\langle 1|\vec{p}|2\rangle = \frac{im}{\hbar}\langle 1|\left[H_0, \vec{r}\right]|2\rangle$$

$$= \frac{im}{\hbar}\left[\langle 1|H_0\vec{r}|2\rangle - \langle 1|\vec{r}H_0|2\rangle\right]$$

$$= \frac{im}{\hbar}\left[\langle 1|E_1\vec{r}|2\rangle - \langle 1|\vec{r}E_2|2\rangle\right]$$

$$= -imw_0\langle 1|\vec{r}|2\rangle$$

가 되어,

$$I_{12} = i\frac{ew_0}{c}A_0^* e^{i(w-w_0)t}\hat{\epsilon}\cdot\langle 1|\vec{r}|2\rangle$$

임을 알 수 있다. 또한

$$\vec{E} = -\frac{\partial}{\partial t}\vec{A}$$

로부터

$$E_0 = iwA_0$$

이고,

$$I_{12} = -\frac{e}{c}E_0^* e^{i(w-w_0)t}\hat{\epsilon}\cdot\langle 1|\vec{r}|2\rangle$$

$$= \frac{V}{2}e^{i(w-w_0)t}$$

이 된다. 여기서

$$V = -\frac{2e}{c}E_0^*\hat{\epsilon}\cdot\langle 1|\vec{r}|2\rangle$$

이다. 따라서 식 (5-1-8)과 (5-1-9)는 다음과 같이 쓸 수 있다.

$$\dot{a}_1 = -\frac{i}{2\hbar}Ve^{i(w-w_0)t}a_2 \qquad (5\text{-}1\text{-}10)$$

$$\dot{a}_2 = -\frac{i}{2\hbar}V^*e^{-i(w-w_0)t}a_1 \qquad (5\text{-}1\text{-}11)$$

라비 진동 _ 원자 속 전자가 두 에너지 준위 사이에서 진동한다

정교수 이제 식 (5-1-10)과 (5-1-11)을 풀어볼 거야.

$t=0$일 때 원자 속 전자의 상태가 $|1\rangle$이라고 하면

$$|\psi(0)\rangle = a_1(0)|1\rangle + a_2(0)|2\rangle = |1\rangle$$

로부터

$$a_1(0) = 1 \qquad (5\text{-}2\text{-}1)$$

$$a_2(0) = 0 \qquad (5\text{-}2\text{-}2)$$

이 된다. 먼저 외부 전자기장의 각진동수 w와 w_0이 같은 경우를 생각해 보자. 즉, $w = w_0$인 경우이다. 이것을 공명 조건이라고 부른다.

이때 식 (5-1-10)과 (5-1-11)은

$$\dot{a}_1 = -\frac{i}{2\hbar} V a_2 \qquad (5-2-3)$$

$$\dot{a}_2 = -\frac{i}{2\hbar} V^* a_1 \qquad (5-2-4)$$

이 된다. 식 (5-2-3)을 미분하면

$$\ddot{a}_1 = -\frac{|V|^2}{4\hbar^2} a_1$$

이다. 이 식을 풀면

$$a_1(t) = \cos\left(\frac{\Omega_0}{2} t\right) \qquad (5-2-5)$$

이고, 식 (5-2-3)으로부터

$$a_2(t) = -i \frac{V^*}{|V|} \sin\left(\frac{\Omega_0}{2} t\right) \qquad (5-2-6)$$

가 된다. 여기서

$$\Omega_0 = \frac{|V|}{\hbar}$$

를 물리학자 라비의 이름을 따서 라비 진동수라고 부른다.

시각 t에서 원자 속 전자가 $|1\rangle$에 있을 확률을 P_1, $|2\rangle$에 있을 확률을 P_2라고 하면

$$P_1 = |a_1|^2 = \cos^2\left(\frac{\Omega_0}{2}t\right) \tag{5-2-7}$$

$$P_2 = |a_2|^2 = \sin^2\left(\frac{\Omega_0}{2}t\right) \tag{5-2-8}$$

이다. 원자 속 전자는 두 에너지 준위 사이에서 진동하는데 이것을 라비 진동이라고 부른다. 다음 그림은 두 확률을 시간에 따라 그린 그래프이다. 검은 선은 P_1을, 파란 선은 P_2를 나타낸다.

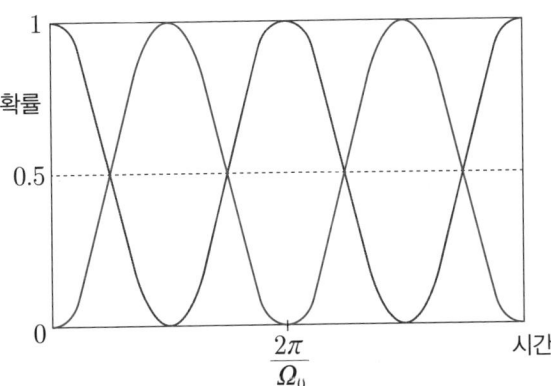

이 그래프를 보면 $t = 0$일 때 $P_1 = 1, P_2 = 0$이다. 즉, 전자는 $|1\rangle$에 있다. 그리고 $t = \dfrac{\pi}{\Omega_0}$일 때 $P_1 = 0, P_2 = 1$이므로 전자는 $|2\rangle$에 있다. 이런 식으로 진동하는 현상을 라비 진동이라고 한다.

물리군 공명 조건을 만족하지 않을 때는 어떻게 되나요?

정교수 공명 조건을 만족하지 않지만 다음 조건을 만족하는 경우를 생각해 볼게.

$$|w - w_0| \ll w_0$$

이때 식 (5-1-10)을 미분하면

$$\ddot{a}_1 = -\frac{i}{2\hbar} V [i(w-w_0) e^{i(w-w_0)t} a_2 + e^{i(w-w_0)t} \dot{a}_2] \qquad (5\text{-}2\text{-}9)$$

이고, 식 (5-1-11)을 식 (5-2-9)에 넣으면

$$\ddot{a}_1 - i(w-w_0)\dot{a}_1 + \frac{|V|^2}{4\hbar^2} a_1 = 0 \qquad (5\text{-}2\text{-}10)$$

이 된다. 같은 방법으로

$$\ddot{a}_2 + i(w-w_0)\dot{a}_2 + \frac{|V|^2}{4\hbar^2} a_2 = 0 \qquad (5\text{-}2\text{-}11)$$

임을 알 수 있다. 초기조건은

$$a_1(0) = 1$$

$$a_2(0) = 0$$

이고

$$\dot{a}_1 = -\frac{i}{2\hbar}Ve^{i(w-w_0)t}a_2$$

에서

$$\dot{a}_1(0) = -\frac{i}{2\hbar}Va_2(0) = 0$$

이고,

$$\dot{a}_2 = -\frac{i}{2\hbar}V^*e^{-i(w-w_0)t}a_1$$

에서

$$\dot{a}_2(0) = -\frac{i}{2\hbar}V^*a_1(0) = -\frac{i}{2\hbar}V^*$$

이다. 이제 식 (5-2-10)과 (5-2-11)을 풀면

$$a_1(t) = e^{i(w-w_0)t/2}\left[\cos\left(\frac{\Omega}{2}t\right) - i\frac{(w-w_0)}{\Omega}\sin\left(\frac{\Omega}{2}t\right)\right]$$

$$a_2(t) = -\frac{i}{\hbar}\frac{V^*}{\Omega}e^{-i(w-w_0)t/2}\sin\left(\frac{\Omega}{2}t\right)$$

가 된다. 여기서 라비 진동수는

$$\Omega = \sqrt{(w-w_0)^2 + \frac{|V|^2}{\hbar^2}}$$

이다. 이때 시각 t에서 원자 속 전자가 $|1\rangle$에 있을 확률을 P_1, $|2\rangle$에 있을 확률을 P_2라고 하면

$$P_1 = |a_1|^2 = \cos^2\left(\frac{\Omega_0}{2}t\right) + \frac{(w-w_0)^2}{\Omega^2}\sin^2\left(\frac{\Omega}{2}t\right)$$

$$P_2 = |a_2|^2 = \frac{|V|^2}{\hbar^2 \Omega^2}\sin^2\left(\frac{\Omega_0}{2}t\right)$$

이다. 그러므로 P_2의 최댓값이 1이 아니다. 즉, 원자 준위의 완전한 반전은 일어나지 않는다.

다음 그림은 $w - w_0 = \frac{|V|}{\hbar}$일 때 P_1과 P_2의 그래프이다. 검은 선은 P_1을, 파란 선은 P_2를 나타낸다.

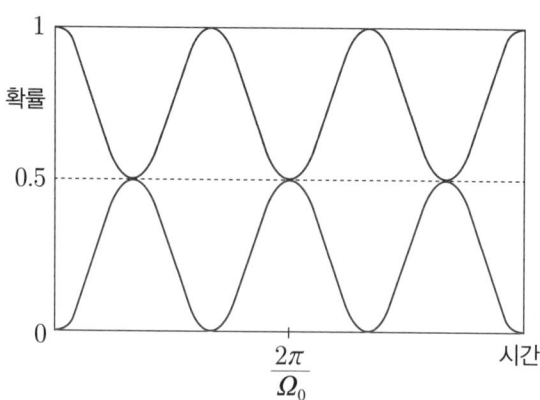

즉, 외부 전자기파가 공명 조건을 만족하면 완전한 반전이 주기적으로 일어나지만, 공명 조건을 만족하지 못하면 완전한 반전이 일어나지 않는 것을 알 수 있다.

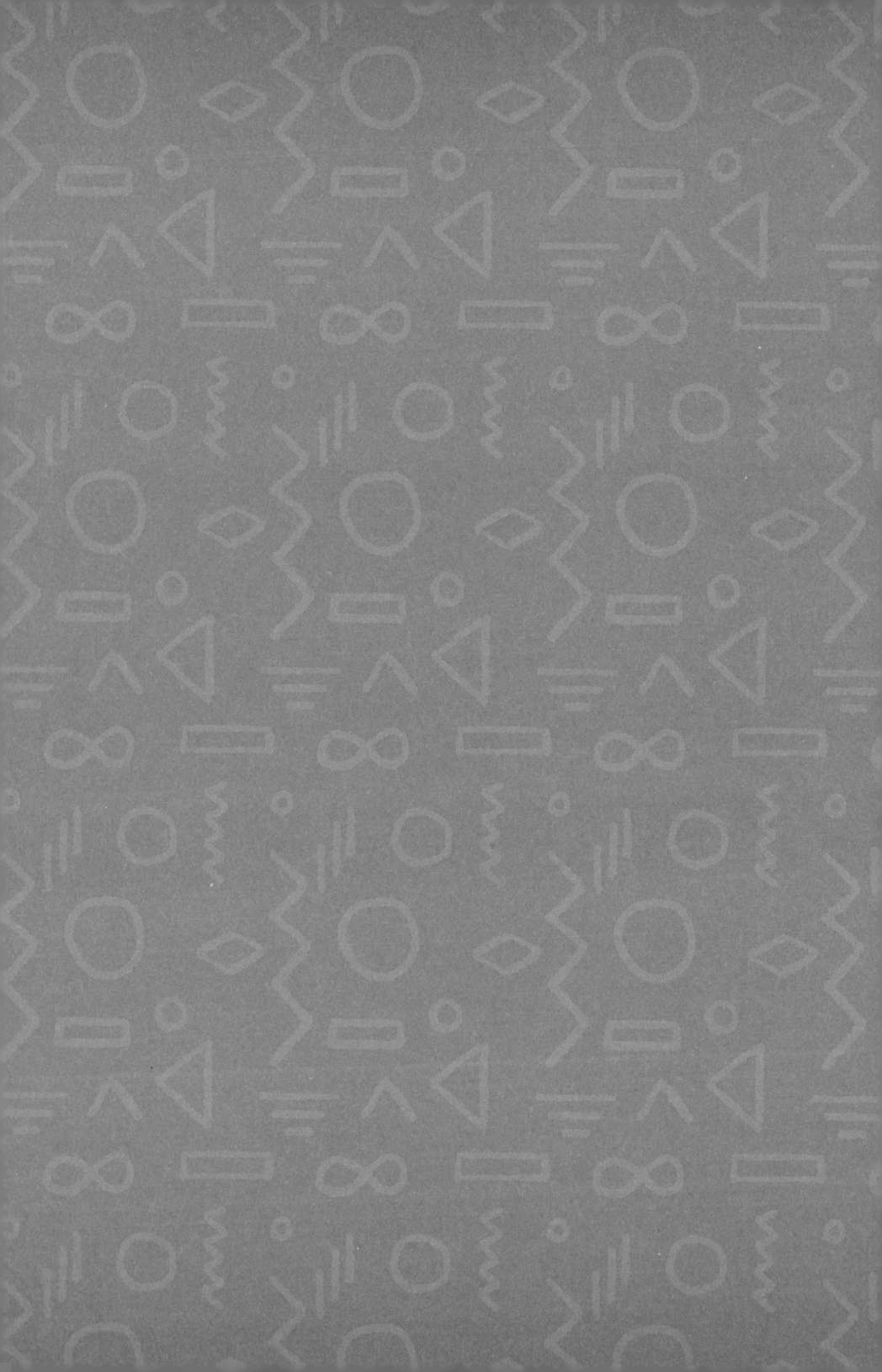

여섯 번째 만남

결맞는 상태

글라우버 _ 로스앨러모스의 최연소 과학자

정교수 그럼 결맞는 상태를 정의하고 반응집광을 이론적으로 예언하여 양자광학의 기초를 확립한 글라우버에 대해 알아볼까?

글라우버(Roy Jay Glauber, 1925~2018, 2005년 노벨 물리학상 수상)

글라우버는 1925년 미국 뉴욕에서 태어났다. 10대 시절부터 과학과 천문학에 관심을 두었던 그는 손수 망원경을 만들었고 심지어 망원경의 거울을 사포로 정밀하게 연마하기까지 했다. 그는 자연사 박물관의 헤이든 천문관에서 강의를 들었다.

1938년 가을, 글라우버는 막 개교한 브롱크스 과학고등학교에 입학해 1941년에 이 학교의 첫 번째 졸업생이 되었다. 그리고 하버드 대학교에서 학부 과정을 밟았다. 2학년을 마친 후에는 맨해튼 프로젝트에 채용되어 18세의 나이에 로스앨러모스 국립 연구소의 최연소 과학자 중 한 명이 되었다. 당시 그는 원자폭탄의 임계 질량을 계산하는 연구를 했다.

로스앨러모스에서 2년을 보내고 하버드 대학교로 돌아온 글라우버는 1946년에 학사 학위를, 1949년에 박사 학위를 받았다. 그의 논문 지도교수는 양자전기역학 연구로 1965년 노벨상을 수상한 이론 물리학자 줄리언 슈윙거(Julian Schwinger)였다.

1956년 글라우버는 영국의 천문학자인 로버트 핸버리 브라운과 리처드 트위스가 실시한 실험에 흥미를 느꼈다. 눈에 보이는 별들의 겉보기 각도 지름을 측정하고자 두 과학자는 서로 일정 거리 떨어진 두 지점에 광 검출기를 설치하였다. 각 검출기는 별빛에서 날아온 광자를 받아들이고, 그 신호를 중앙 장치로 전송하였다. 이 중앙 장치는 두 검출기에 빛이 같은 순간에 도착하는 경우가 얼마나 자주 일어나는지를 조사해, 별빛의 퍼짐 정도를 알아냈다. 글라우버는 이 과정을 전자기파의 양자화로 풀어냈다. 그의 연구는 1963년 〈Coherent and Incoherent States of the Radiation Field〉라는 제목의 기념비적인 논문에 집대성되었고, 그는 이 논문으로 노벨 물리학상을 수상했다.

광자가 어떻게 상호작용 하는지에 대한 글라우버의 이론은 디랙의 이론과 일부 모순되었다. 하지만 실험으로 글라우버의 이론이 옳다는 것이 확인되었다. 글라우버는 여러 종류의 진동수를 가진 전자기파로 구성된 백색광과 일정한 진동수의 전자기파를 내는 레이저 광의 차이를 설명할 수 있었다. 그의 이론은 훗날 양자컴퓨터와 양자암호를 개발하는 과정에서 쓰였다.

결맞는 상태 _ 빛의 선명도가 최대가 되다

정교수   글라우버는 빛을 이루는 양자인 광자를 통해 결맞음을 연구하고, 결맞는 상태를 찾아낸 연구로 노벨 물리학상을 수상했지.

물리군   결맞음, 결맞는 상태가 뭐죠?

정교수   빛을 광자라는 아주 작은 입자들의 흐름으로 생각할 수 있어. 이 광자들이 질서 있게 일정한 간격으로 같은 방향과 속도로 움직이는 상태가 결맞는 상태이고, 이러한 성질이 결맞음이야. 마치 군인들이 박자를 맞춰 행진하는 모습처럼, 광자들이 제각각이 아니라 딱 맞춘 듯 움직이는 거지. 보통 손전등의 빛은 광자들이 제멋대로 나와. 즉, 방향도 다르고 위상도 들쭉날쭉해서 결맞음이 약한 상태의 빛이돼. 그런데 레이저 속의 빛은 광자들이 한 방향으로 정렬된 형태로 나와. 그러니까 레이저 속의 빛은 결맞는 상태가 되지.

물리군   그렇군요.

정교수   글라우버는 영의 이중 슬릿 실험을 다시 들여다보았어.

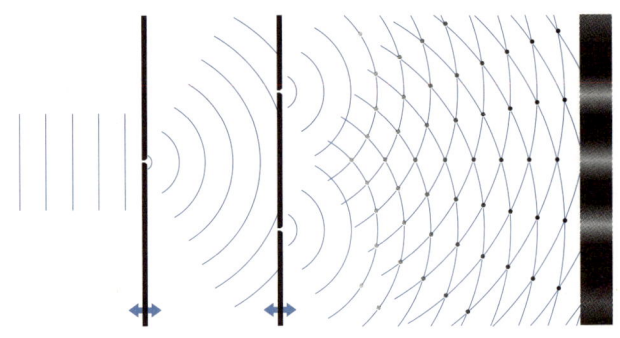

슬릿을 통과한 빛의 간섭이 잘 일어나면 밝고 어두운 무늬를 더욱 선명하게 볼 수 있다. 글라우버는 이 선명도가 최대인 빛을 빛의 결맞는 상태라고 불렀다. 그리고 이 상태는 네 번째 만남 6절에서 도입한 두 연산자 $\hat{x}$와 $\hat{p}$의 불확정성이 최소인 경우라는 것을 알아냈다. 식 (4-6-7)로부터

$$\hat{x} = \sqrt{\frac{\hbar}{2w}}(\hat{a} + \hat{a}^\dagger)$$

$$\hat{p} = i\sqrt{\frac{\hbar w}{2}}(\hat{a}^\dagger - \hat{a})$$

이다. 이때

$$[\hat{a}, \hat{a}^\dagger] = 1$$

또는

$$[\hat{x}, \hat{p}] = i\hbar$$

가 된다.

양자광학의 최소 불확정성원리는

$$(\Delta x)(\Delta p) = \frac{\hbar}{2} \qquad (6\text{-}2\text{-}1)$$

로 주어진다. 이때 최소 불확정성원리가 성립하는 상태인 결맞는 상태를 $|z\rangle$라고 하면

$$(\Delta x)^2 = \langle z | (\hat{x} - \langle \hat{x} \rangle)^2 | z \rangle \qquad (6\text{-}2\text{-}2)$$

$$(\Delta p)^2 = \langle z | (\hat{p} - \langle \hat{p} \rangle)^2 | z \rangle \qquad (6\text{-}2\text{-}3)$$

로 정의한다.

여기서 식 (6-2-2)와 (6-2-3)은 각각

$$(\Delta x)^2 = \langle z | (\hat{x} - \langle \hat{x} \rangle)(\hat{x} - \langle \hat{x} \rangle) | z \rangle$$

$$= \| (\hat{x} - \langle \hat{x} \rangle) | z \rangle \|^2$$

$$(\Delta p)^2 = \langle z | (\hat{p} - \langle \hat{p} \rangle)(\hat{p} - \langle \hat{p} \rangle) | z \rangle$$

$$= \| (\hat{p} - \langle \hat{p} \rangle) | z \rangle \|^2$$

이 된다.

이제

$$|f\rangle = (\hat{x} - \langle \hat{x} \rangle) | z \rangle \qquad (6\text{-}2\text{-}4)$$

$$|g\rangle = (\hat{p} - \langle \hat{p} \rangle) | z \rangle \qquad (6\text{-}2\text{-}5)$$

라고 두면

$$(\Delta x)^2 = \langle f | f \rangle$$

$$(\Delta p)^2 = \langle g | g \rangle$$

로 쓸 수 있다. 이때 다음과 같이 놓자.

$$|X\rangle = |f\rangle - \frac{\langle g|f\rangle}{\langle g|g\rangle}|g\rangle \qquad (6\text{-}2\text{-}6)$$

그러면

$$\||X\rangle\|^2 = \langle X|X\rangle \geq 0$$

이다. 따라서

$$\left(\langle f| - \frac{\langle f|g\rangle}{\langle g|g\rangle}\langle g|\right)\left(|f\rangle - \frac{\langle g|f\rangle}{\langle g|g\rangle}|g\rangle\right) \geq 0 \qquad (6\text{-}2\text{-}7)$$

또는

$$\langle f|f\rangle - \frac{\langle g|f\rangle\langle f|g\rangle}{\langle g|g\rangle} \geq 0 \qquad (6\text{-}2\text{-}8)$$

이 된다. 즉, 다음과 같다.

$$\langle f|f\rangle\langle g|g\rangle - \langle g|f\rangle\langle f|g\rangle \geq 0 \qquad (6\text{-}2\text{-}9)$$

한편 $\langle f|g\rangle = \langle g|f\rangle^*$로부터

$$\langle f|f\rangle\langle g|g\rangle \geq |\langle f|g\rangle|^2 \qquad (6\text{-}2\text{-}10)$$

임을 알 수 있다. 또한

$$\langle f | g \rangle = \langle z | (\hat{x} - \langle \hat{x} \rangle)(\hat{p} - \langle \hat{p} \rangle) | z \rangle$$

$$= \langle z | \hat{x}\hat{p} | z \rangle - \langle \hat{x} \rangle \langle \hat{p} \rangle$$

이다. 여기서

$$\hat{x}\hat{p} = \frac{1}{2}(\hat{x}\hat{p} + \hat{p}\hat{x}) + \frac{1}{2}(\hat{x}\hat{p} - \hat{p}\hat{x})$$

$$= \frac{1}{2}(\hat{x}\hat{p} + \hat{p}\hat{x}) + \frac{i\hbar}{2}$$

이므로

$$\langle f | g \rangle = \frac{1}{2} \langle z | \hat{x}\hat{p} + \hat{p}\hat{x} | z \rangle - \langle \hat{x} \rangle \langle \hat{p} \rangle + \frac{i\hbar}{2}$$

이다. $\frac{1}{2} \langle z | \hat{x}\hat{p} + \hat{p}\hat{x} | z \rangle$는 실수이므로

$$|\langle f | g \rangle|^2 = \left(\frac{1}{2} \langle z | \hat{x}\hat{p} + \hat{p}\hat{x} | z \rangle - \langle \hat{x} \rangle \langle \hat{p} \rangle \right)^2 + \frac{\hbar^2}{4} \geq \frac{\hbar^2}{4} \quad (6\text{-}2\text{-}11)$$

이다. 따라서 식 (6-2-10)으로부터

$$(\Delta x)^2 (\Delta p)^2 \geq \frac{\hbar^2}{4}$$

이고, 최소 불확정성을 만족하는 경우에 등호가 성립한다. 이것은 식 (6-2-9)에서

$$|f\rangle = -i\lambda |g\rangle \quad (6\text{-}2\text{-}12)$$

일 때이다. 여기서 λ는 실수이다.

물리군 왜 그런 거죠?

정교수 식 (6-2-12)에서

$$\langle f | = i\lambda \langle g | \qquad (6\text{-}2\text{-}13)$$

이지? 식 (6-2-12)와 (6-2-13)을 식 (6-2-9)에 넣으면

$$\langle f | f \rangle \langle g | g \rangle - \langle g | f \rangle \langle f | g \rangle = 0$$

이 되거든.

따라서 최소 불확정성 상태 $|z\rangle$는

$$(\hat{x} - \langle \hat{x} \rangle) | z \rangle = -i\lambda (\hat{p} - \langle \hat{p} \rangle) | z \rangle$$

또는

$$(\hat{x} + i\lambda \hat{p}) | z \rangle = (\langle \hat{x} \rangle + i\lambda \langle \hat{p} \rangle) | z \rangle \qquad (6\text{-}2\text{-}14)$$

가 된다. 양자광학에서 광자의 에너지 연산자

$$\hat{H} = \frac{1}{2}\hat{p}^2 + \frac{1}{2}w^2 \hat{x}^2$$

을 떠올리자. $\lambda = \dfrac{1}{w}$로 택하고, 식 (4-6-7)을 이용하면 식 (6-2-14)는

$$\hat{a}|z\rangle = \sqrt{\frac{w}{2\hbar}}\left(\langle\hat{x}\rangle + i\frac{\langle\hat{p}\rangle}{w}\right)|z\rangle$$

가 된다. 이때 복소수

$$z = \sqrt{\frac{w}{2\hbar}}\left(\langle\hat{x}\rangle + i\frac{\langle\hat{p}\rangle}{w}\right)$$

를 도입하면 결맞는 상태는

$$\hat{a}|z\rangle = z|z\rangle$$

를 만족한다.

결맞는 상태의 성질 _ 광자수의 확률, 기댓값, 표준편차

정교수   빛이 결맞는 상태에 있을 때 그 빛을 결맞는 상태 빛이라고 불러. 이제 결맞는 상태를 광자수 상태의 중첩으로 나타내 볼게. 그 과정은 다음과 같아.

$$|z\rangle = \sum_{n=0}^{\infty} c_n |n\rangle \qquad (6\text{-}3\text{-}1)$$

물리군 c_n은 어떻게 구하나요?

정교수   결맞는 상태의 정의를 이용하면 돼.

$$\hat{a}\,|\,z\,\rangle = \hat{a}\sum_{n=0}^{\infty} c_n\,|\,n\,\rangle$$

$$= \sum_{n=0}^{\infty} c_n \hat{a}\,|\,n\,\rangle$$

$$= \sum_{n=1}^{\infty} c_n \sqrt{n}\,|\,n-1\,\rangle \tag{6-3-2}$$

물리군 왜 합이 1부터 시작하죠?

정교수 $\hat{a}\,|\,0\,\rangle = 0$이기 때문이야. 식 (6-3-2)는

$$\hat{a}\,|\,z\,\rangle = c_1\sqrt{1}\,|\,0\,\rangle + c_2\sqrt{2}\,|\,1\,\rangle + c_3\sqrt{3}\,|\,2\,\rangle + \cdots$$

$$= \sum_{n=0}^{\infty} c_{n+1}\sqrt{n+1}\,|\,n\,\rangle$$

으로 쓸 수 있어. 그러니까

$$\sum_{n=0}^{\infty} c_{n+1}\sqrt{n+1}\,|\,n\,\rangle = z\sum_{n=0}^{\infty} c_n\,|\,n\,\rangle$$

이지. 따라서

$$c_{n+1}\sqrt{n+1} = zc_n$$

또는

$$c_{n+1} = \frac{z}{\sqrt{n+1}} c_n$$

이 돼. 이 식에 $n = 0, 1, 2, \cdots$ 를 차례로 대입하면

$$c_1 = \frac{z}{\sqrt{1}} c_0$$

$$c_2 = \frac{z}{\sqrt{2}} c_1$$

$$c_3 = \frac{z}{\sqrt{3}} c_2$$
$$\vdots$$

가 되지. 그러니까

$$c_n = \frac{z^n}{\sqrt{n!}} c_0$$

이야. 즉, 결맞는 상태는 다음과 같아.

$$|z\rangle = c_0 \sum_{n=0}^{\infty} \frac{z^n}{\sqrt{n!}} |n\rangle \qquad (6\text{-}3\text{-}3)$$

물리군 c_0은 어떻게 구하죠?

정교수 우리는 모든 상태의 크기가 1이기를 원해.

결맞는 상태의 크기가 1이려면

$$\langle z | z \rangle = 1$$

이어야 한다. c_0을 실수라고 하면

$$\langle z | = c_0 \sum_{n=0}^{\infty} \frac{(z^*)^n}{\sqrt{n!}} \langle n |$$

이므로

$$\langle z | z \rangle = c_0^2 \sum_{n=0}^{\infty} \frac{(z^*)^n z^n}{n!}$$

이다. 이때 $|z|^2 = z^*z$이니까

$$c_0^2 \sum_{n=0}^{\infty} \frac{|z|^{2n}}{n!} = 1$$

이 된다. 여기서

$$e^x = \sum_{n=0}^{\infty} \frac{x^n}{n!}$$

을 이용하자. 참고로 이 식의 증명은 이 시리즈의 《양자혁명》에 있다. 따라서

$$c_0^2 e^{|z|^2} = 1$$

이므로

$$c_0 = e^{-\frac{1}{2}|z|^2}$$

이 된다. 그러니까 결맞는 상태는

$$|z\rangle = e^{-\frac{1}{2}|z|^2} \sum_{n=0}^{\infty} \frac{z^n}{\sqrt{n!}} |n\rangle \qquad (6\text{-}3\text{-}4)$$

이다.

이제 결맞는 상태 $|z\rangle$ 속에서 n개의 광자를 발견할 확률을 P_n이라고 하면

$$P_n = |c_n|^2$$

이고,

$$c_n = \langle n | z \rangle$$

이다. 그러므로 다음과 같다.

$$\begin{aligned} P_n &= |\langle n | z \rangle|^2 \\ &= \left| e^{-\frac{1}{2}|z|^2} \frac{z^n}{\sqrt{n!}} \right|^2 \\ &= e^{-|z|^2} \frac{|z|^{2n}}{n!} \end{aligned} \qquad (6\text{-}3\text{-}5)$$

또한 결맞는 상태의 빛에서 광자수의 기댓값은

$$\langle n \rangle = \langle z | \hat{N} | z \rangle$$

$$= e^{-|z|^2} \sum_{n=0}^{\infty} \frac{(z^*)^n}{\sqrt{n!}} \langle n | \hat{N} \sum_{m=0}^{\infty} \frac{z^m}{\sqrt{m!}} | m \rangle$$

$$= e^{-|z|^2} \sum_{n=0}^{\infty} \frac{(z^*)^n}{\sqrt{n!}} \langle n | \sum_{m=0}^{\infty} \frac{z^m}{\sqrt{m!}} m | m \rangle$$

$$= e^{-|z|^2} \sum_{n=0}^{\infty} \frac{(z^*)^n}{\sqrt{n!}} \sum_{m=0}^{\infty} \frac{z^m}{\sqrt{m!}} m \langle n | m \rangle$$

$$= e^{-|z|^2} \sum_{n=0}^{\infty} \frac{(z^*)^n}{\sqrt{n!}} \sum_{m=0}^{\infty} \frac{z^m}{\sqrt{m!}} m \delta_{nm}$$

$$= e^{-|z|^2} \sum_{n=0}^{\infty} \frac{(|z|)^{2n}}{n!} n$$

$$= e^{-|z|^2} \sum_{n=1}^{\infty} \frac{(|z|)^{2n}}{(n-1)!}$$

$$= |z|^2 e^{-|z|^2} \sum_{n=1}^{\infty} \frac{(|z|)^{2(n-1)}}{(n-1)!}$$

$$= |z|^2 e^{-|z|^2} \sum_{n=0}^{\infty} \frac{(|z|)^{2n}}{n!}$$

$$= |z|^2 e^{-|z|^2} e^{|z|^2}$$

$$= |z|^2 \qquad (6\text{-}3\text{-}6)$$

이다. 같은 방법으로

$$\langle n^2 \rangle = \langle z | \hat{N}^2 | z \rangle = |z|^2 + |z|^4 \tag{6-3-7}$$

이 된다. 따라서 광자수의 표준편차(불확정성)를 Δn이라고 하면

$$\Delta n = |z| = \sqrt{\langle n \rangle} \tag{6-3-8}$$

이고, 식 (6-3-5)는

$$P_m = e^{-\langle n \rangle} \frac{\langle n \rangle^m}{m!} \tag{6-3-9}$$

으로 쓸 수 있다. 이것은 수학자 푸아송이 알아낸 푸아송 분포이다. 다음 그림은 $\langle n \rangle = 10$일 때의 그래프이다.

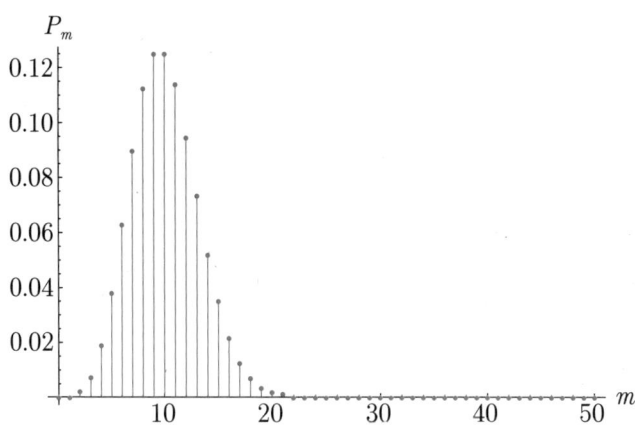

다음 그림은 $\langle n \rangle = 20$일 때의 그래프이다.

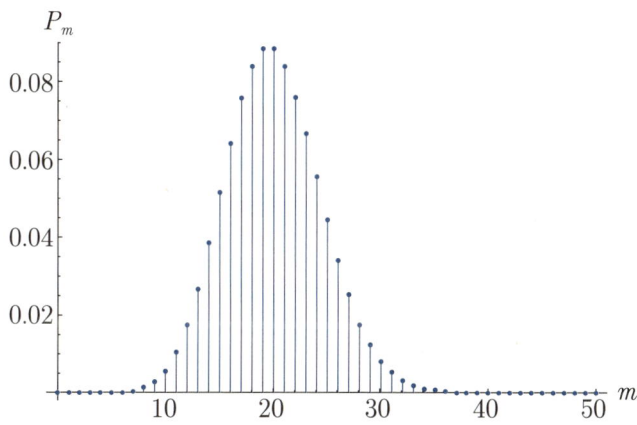

결맞는 상태에 있는 빛의 총에너지는 다음과 같이 주어진다.

$$\langle z | \hat{H} | z \rangle = \hbar w \left(\langle n \rangle + \frac{1}{2} \right)$$

반응집광 _ 방출되는 광자가 분리되는 현상

물리군 글라우버가 반응집광을 이론적으로 예언했다던데요. 반응집광은 뭔가요?

정교수 양자광의 대표적인 특징 중 하나야. 광자의 반응집성(Photon antibunching)은 광원에서 방출될 때 개별 광자가 일시적으로 분리되는 양자역학적 현상을 말해. 이렇게 방출되는 빛을 반응집광이라 하고, 반응집광과 반대로 광자들이 몰려다니는 행동을 보이면 응집광이라고 불러.

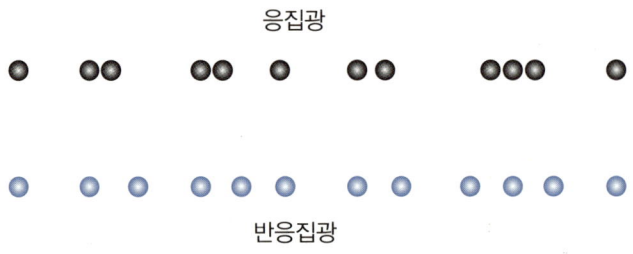

광자를 덩어리로 방출하거나 함께 '뭉쳐' 방출하는 기존 광원과 달리, 반응집광은 어떤 시간 간격 동안 한 번에 하나씩 감지되지. 이러한 현상은 양자광학을 통해서만 설명할 수 있으며 고급 양자 기술의 기반을 제공한다네.

일곱 번째 만남

양자광 기술

애슈킨의 광학 핀셋 _레이저 빔으로 미세한 물체를 붙잡다

정교수 이번에는 광학 핀셋을 발명한 애슈킨을 소개할게.

애슈킨(Arthur Ashkin, 1922~2020, 2018년 노벨 물리학상 수상, 사진 출처: Bengt Nyman/Wikimedia Commons)

애슈킨은 1922년 미국 뉴욕 브루클린의 우크라이나계 유대인 가정에서 태어났다. 그는 1940년 브루클린의 제임스 매디슨 고등학교를 졸업했다. 그 후 컬럼비아 대학교에 다녔으며, 컬럼비아 방사선 연구소의 기술자로 일하면서 미군 레이더 시스템용 마그네트론을 제작하는 임무를 맡았다. 1947년에 애슈킨은 컬럼비아 대학교에서 물리학 학사 학위를 받았다. 그리고 코넬 대학교에 입학하여 핵물리학을 공부했다.

코넬 대학교

1952년 애슈킨은 코넬 대학교에서 박사 학위를 받은 뒤, 컬럼비아 대학교의 지도교수였던 시드니 밀먼의 추천으로 벨 연구소에서 일하기 시작했다.

애슈킨은 벨 연구소에서 1960년부터 1961년까지 마이크로파 분야에서 일하다가 레이저 연구로 전환했다. 당시 그는 비선형광학, 광섬유, 파라메트릭 발진기 및 파라메트릭 증폭기를 연구했고, 1960년대에 압전 결정의 광굴절 효과를 발견했다.

물리군 애슈킨은 어떤 업적으로 노벨 물리학상을 받았나요?

정교수 그의 노벨상 수상 이유는 광학 핀셋 발명이었어.

물리군 광학 핀셋이 뭐죠?

정교수 고도로 집중된 레이저 빔을 사용하여 핀셋과 유사한 방식으로 원자나 나노 입자 및 물방울과 같은 미세한 물체를 붙잡거나 이동시키는 장치야.

물리군 나노 입자는 뭔가요?

정교수 지름이 1 내지 100나노미터인 입자를 말해. 1나노미터는 십억분의 일 미터이지.

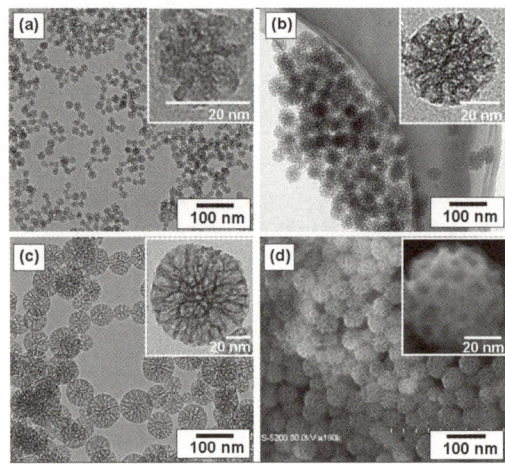

나노 입자 사진

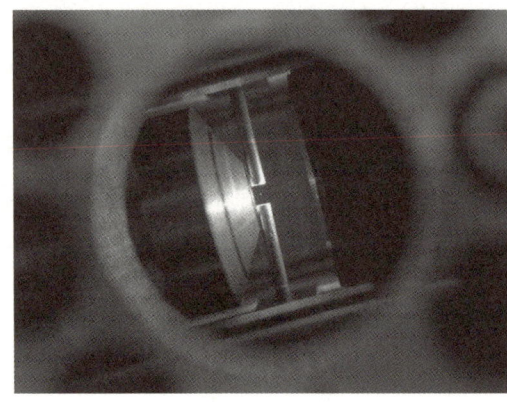

광학 핀셋에 갇힌 나노 입자(지름 103나노미터)의 사진. 나노 입자는 중간에 있는 작은 밝은 점이다.(출처: Bjschellenberg/ Wikimedia Commons)

광학 핀셋은 광선이 물체로부터 반사되거나 물체에 흡수될 때 물체 표면에 미치는 압력을 이용하는데, 이 압력을 방사압력이라고 부른다. 일반 광은 넓게 퍼지기 때문에 방사압력이 작지만 레이저 광은 한 점에 모이는 성질이 있어 큰 방사압력을 줄 수 있다.

나노 입자에 레이저 빔을 쪼이면 입자에 작용한 방사압력이 입자를 포획하는 힘을 만들어 낸다. 이를 통해 단일 레이저 빔은 한 쌍의 핀셋처럼 작동하여 개별적인 원자, 분자, 생물학적 세포들을 미세하게 조작할 수 있다.

1970년 벨 연구소의 애슈킨은 이 방법으로 나노 입자를 붙잡거나 이동시킬 수 있었다. 그는 이 연구를 지속하여 1986년에 동료들과 함께 실제 사용 가능한 광학 핀셋 아이디어를 논문으로 발표했다.

1980년대 후반, 애슈킨과 지드직(Joseph M. Dziedzic)은 광학 핀셋 기술을 생물학에 최초로 적용해 담배 모자이크 바이러스와 대장균 박테리아를 포획하는 데 성공했다.

광학 핀셋은 생물학의 다른 영역에서도 유용한 것으로 입증되었다. 유전과 염색체 구조 연구에 널리 쓰였고, 2003년에는 세포 분류 분야에도 적용되었다.

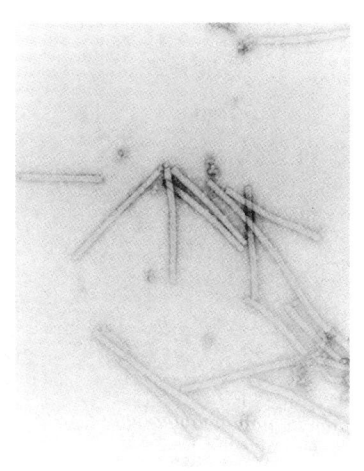

담배 모자이크 바이러스

주파수 빗의 발명 _ 일정한 주파수 간격의 날카로운 스펙트럼선

정교수 이제 주파수 빗을 발명한 과학자들의 이야기를 해보려고 해. 먼저 미국의 물리학자 홀에 대해 살펴볼게.

홀(John Lewis Hall, 1934~2005년 노벨 물리학상 수상)

홀은 미국 콜로라도주 덴버에서 태어났다. 그는 카네기 공과대학에서 1956년에 학사, 1958년에 석사, 1961년에 박사 학위를 취득했다. 1962년부터 2004년 은퇴할 때까지는 미국 상무부 산하 국립표준국(National Bureau of Standards, 현 NIST)에서 근무했다.

NIST

두 번째로 소개할 과학자는 독일의 헨슈이다.

헨슈는 독일 하이델베르크 헬름홀츠 김나지움에서 중고등학교 교육을 받았고, 1960년대에 하이델베르크 대학교에서 박사 학위를 받았다. 1970년부터 1972년까지는 미국 스탠퍼드 대학교에서 박사후 연구원으로 일했다. 그는 1975년에 스탠퍼드 대학교의 조교수가 되었고, 독일로 돌아와 막스 플랑크 양자광학 연구소를 이끌었다.

헨슈(Theodor W. Hänsch, 1941~, 2005년 노벨 물리학상 수상, 사진 출처: Markus Pössel/Wikimedia Commons)

막스 플랑크 양자광학 연구소(출처: Ghost writ0r/Wikimedia Commons)

1970년 헨슈는 매우 높은 스펙트럼 분해능을 생성하는 새로운 유형의 레이저를 발명했다. 이것을 가지고 그는 이전보다 정밀도를 높여 발머 계열의 주파수를 보다 정확하게 측정하는 데 성공했다. 1990년대 후반에 그와 동료들은 광학 주파수 빗 생성기라는 장치를 사용하여 레이저 광의 주파수를 훨씬 더 높은 정밀도로 측정하는 새로운 방법을 개발했다. 주파수 빗 생성기의 발명으로 홀과 헨슈는 노벨 물리학상을 공동 수상했다.

물리군　주파수 빗 생성기가 뭐예요?

정교수　일정한 시간 간격으로 파장이 아주 짧은 전자기파(극초단파)를 생성하는 장치야. 이 장치를 이용하면 일정한 주파수 간격을 가진 수십만 개의 날카로운 스펙트럼선을 만들 수 있어. 이 모습이 빗을 닮아 주파수 빗이라고 부르지.

주파수 빗에서 나오는 빛의 스펙트럼
(출처: ESO)

주파수 빗은 현재 전 세계 수많은 실험실에서 광학 주파수 측정의 기초로 쓰인다. 헨슈는 최초로 좁은 선폭 조정 가능 레이저를 생산했다. 이는 더욱 좁은 선폭 다중 프리즘 격자 레이저 발진기의 개발에 큰 영향을 미쳤다.

주파수 빗 레이저 분광계로 현장에서 미량 가스를 감지하는 모습을 그린 그림

레이저 냉각 _ 원자의 속력을 느리게 만들다

정교수 이번에는 레이저 냉각 기술로 노벨 물리학상을 받은 세 사람을 만나볼까? 첫 번째로 등장하는 물리학자는 필립스야.

필립스(William Daniel Phillips, 1948~, 1997년 노벨 물리학상 수상, 사진 출처: Markus Pössel/Wikimedia Commons)

필립스는 미국 펜실베이니아주 윌크스배러에서 태어났다. 그의 어머니는 이탈리아 혈통이고 아버지는 웨일스 혈통이다. 필립스는 윌크스배러에서 고등학교를 다녔고 1966년에 수석으로 졸업했다. 그는 1970년 주니아타 칼리지를 최우등으로 졸업했다. 그 후 매사추세츠 공과대학(MIT)에서 물리학 박사 학위를 받고 1978년에 미국 국립표준국에 입사했다.

1997년 필립스는 미국 국립표준기술연구소(NIST)에서 기체 원자의 움직임을 늦추는 레이저 냉각 기술을 발견했다. 이 업적으로 그는 클로드 코엔타누지, 스티븐 추와 함께 노벨 물리학상을 수상했다.

두 번째로 소개할 과학자는 프랑스의 코엔타누지이다.

코엔타누지(Claude Cohen-Tannoudji, 1933~, 1997년 노벨 물리학상 수상, 사진 출처: Amir Bernat/Wikimedia Commons)

 코엔타누지는 프랑스령 알제리의 콩스탕틴에서 태어났다. 그의 부모는 알제리계 유대인이었다. 1953년 알제에서 중등학교를 마친 코엔타누지는 고등사범학교에 입학하기 위해 파리로 떠났다. 1962년 말에 그는 박사 학위를 받았다.

 1964년에 파리 대학의 교수가 된 코엔타누지는 양자역학을 강의하면서 원자-광자 상호작용에 대해 연구했다. 1973년 그는 콜레주 드 프랑스의 교수가 되었다. 1980년대 초, 그는 레이저 빔이 원자에 작용하는 방사압력을 연구했고, 알랭 아스페(Alain Aspect, 2022년 노벨 물리학상 수상), 크리스토프 살로몽(Christophe Salomon), 장 달리바르(Jean Dalibard)와 함께 레이저 냉각 및 트래핑을 연구하기 위해 연구소를 설립했다.

 세 번째로 만나볼 과학자는 스티븐 추다.

추(Steven Chu, 1948~, 1997년 노벨 물리학상 수상)

추는 1948년 미국 미주리주 세인트루이스에서 태어났으며 중국계 혈통이다. 그는 뉴욕주 가든시티에 있는 가든시티 고등학교에 다녔다. 1970년 로체스터 대학교에서 수학과 물리학 학사 학위를, 1976년 버클리 대학교에서 물리학 박사 학위를 받았다.

박사 학위를 취득한 후 추는 버클리에서 박사후 연구원으로 2년 동안 근무했다. 그리고 벨 연구소에 합류하여 레이저 냉각 연구를 수행했다. 이후 벨 연구소를 떠나 1987년 스탠퍼드 대학교의 물리학 교수가 되었다. 2004년 8월에는 미국 에너지부 산하 국립연구소인 로런스 버클리 국립연구소 소장으로 임명되었고, 버클리 대학교의 물리학과 및 분자세포생물학과 교수가 되었다. 추 교수의 지휘로 로런스 버클리 국립연구소는 바이오 연료와 태양에너지 연구의 중심지로 자리매김했다.

추의 초기 연구는 레이저 냉각 기술이었다. 그와 벨 연구소의 동료들은 쌍으로 마주 보는 6개의 레이저 빔을 가지고 원자를 냉각하는 방법을 개발했어. 이 기술은 매우 정밀한 원자시계를 구성하는 데 쓰였어.

물리군 레이저 빔으로 어떻게 온도를 낮출 수 있나요?

정교수 온도는 운동에너지의 평균에 의해 결정돼. 운동에너지는 원자들의 평균속력의 제곱에 비례하니까, 원자들의 속력이 느려지면 원자들로 이루어진 계의 온도가 낮아지지. 원자들로 이루어진 계에 레이저 빔을 쪼이면 원자는 레이저 빔을 이루는 광자를 흡수하고 방출하는 과정을 반복해. 이 과정에서 원자의 평균속력은 줄어들고 이로 인해 원자들로 이루어진 계의 온도가 낮아지는 거야.

레이저 냉각 리튬 원자의 사진
(출처: Darth luigi/Wikimedia Commons)

메타물질 _ 투명 망토의 비밀

정교수 이번에는 메타물질에 대해 이야기할게.

물리군 메타물질이 뭐예요?

정교수 메타물질은 자연에 있는 보통의 물질에 없는 특성을 가진 인공적인 물질을 말해. 주로 전자기파와 관련된 광학적 특성을 가진 것이 많지만, 음파나 열전도 등 다른 여러 가지 파동에 대응하는 메타물질도 있어.

물리군 메타물질은 어디에 이용되나요?

정교수 메타물질의 활용 분야는 매우 넓어. 층간 소음을 비롯한 생활 소음을 줄이거나 음파 탐지기에 잡히지 않는 스텔스 잠수함에서, 또한 투명 망토나 고성능 렌즈, 소형 안테나 등에 사용할 수 있지.

물리군 메타물질의 재미있는 성질을 알려주세요.

정교수 빛이 물질 속으로 들어가면 굴절돼. 일반적인 물질은 양의 굴절률을 가져. 그러니까 일반적인 물질인 물속에 젓가락을 넣으면 다음과 같이 굴절되지.

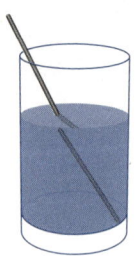

하지만 만일 컵에 물 대신 특별한 메타물질을 넣으면 젓가락이 다음 그림과 같이 반대 방향으로 꺾여 보일 수 있어.

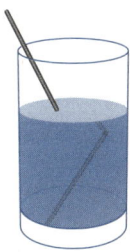

이때 이 메타물질은 음의 굴절률을 가졌다고 말해. 1967년 베셀라고가 음의 굴절률이 이론적으로 가능한 것을 처음 알아냈지.

베셀라고(Victor Georgievich Veselago, 1929~2018, 사진 출처: Guérin Nicolas/Wikimedia Commons)

베셀라고는 고등학교 졸업반 때 아마추어 무선통신인 햄 라디오에 취미가 있었다. 그는 이를 통해 전기와 관련된 물리학에 관심을 가지게 되어 모스크바 주립대학교 물리학과에 입학했다.

모스크바 주립대학교(출처: I.s.kopytov/Wikimedia Commons)

대학 시절에도 베셀라고는 여전히 무선에 관심을 두었고 햄을 이용한 무선통신을 즐겼다. 그는 대학 졸업 후 라디오 천문학의 대가인 카이킨(Semen Emmanuilovich Khaikin) 교수를 만나러 레베데프 연구소로 갔다. 그곳에서 카이킨 교수에게 방사선 천문학을 배웠다. 그 후 1967년 베셀라고는 음의 굴절률이 이론적으로 가능하다는 논문을 발표했다. 이것은 메타물질의 이론적 가능성을 시사하는 첫 논문이었다.

이번에는 메타물질로 유명한 과학자 펜드리에 대해 알아보자.

펜드리(John Brian Pendry, 1943~,
사진 출처: Per Henning/NTNU/Wikimedia Commons)

펜드리는 영국 맨체스터에서 태어났다. 그는 케임브리지 다우닝 칼리지에서 자연과학 학위를 받은 후 1969년에서 1975년 사이에 연구원으로 임명되었다. 1972년과 1973년에는 벨 연구소에서 광전자 분광법을 연구했고, 1975년부터 1981년까지 SERC Daresbury Laboratory의 이론 그룹 책임자로 지내면서 임페리얼 칼리지 런던의 이론 물리 교수로 일했다.

임페리얼 칼리지 런던에서 펜드리는 영국의 선도적인 이론 표면 물리학자로서의 위치를 유지했다. 동시에 무질서한 매질에서 전자의 거동을 연구하여 일반적인 1차원 산란 문제에 대한 완전한 해결책을 도출했다. 1994년에 그는 빛과 금속 시스템의 상호작용을 발견하여 광자 띠 구조에 대한 첫 번째 논문을 발표했다. 이 논문을 통해 그는 메타물질의 아이디어를 얻었다.

2000년 펜드리는 베셀라고의 음의 굴절률 연구를 확장하여, 이론적으로 완벽한 초점을 가진 렌즈를 만드는 간단한 방법을 제안했다. 이 논문에서 언급한 음의 굴절률을 가진 완벽한 렌즈를 슈퍼 렌즈라고 부른다. 펜드리의 아이디어는 실험으로 확인되었고, 슈퍼 렌즈 개념은 나노 스케일 광학에 혁명을 일으켰다.

2006년 펜드리는 빛을 굴절시켜 물체를 보이지 않게 하는 아이디어를 고안했다. 일반적으로 투명 망토로 알려진 이 아이디어는 메타 물질 분야의 최근 연구에 많은 자극을 주었다.

물리군   투명 망토를 쓰면 정말 안 보이나요?
정교수   물론이야.

(출처: Z22/Wikimedia Commons)

물리군   어떻게 이런 현상이 가능한 거죠?

정교수   물체에 빛을 쪼이면 물체에서 빛이 반사되어 그 빛이 눈에 들어가서 물체가 보이는 거야. 하지만 메타물질로 만든 투명 망토를 물체에 씌우면 빛이 메타물질을 휘어 감듯이 꺾여서 지나가거든. 그러니까 물체에 빛이 도달하지 않아.

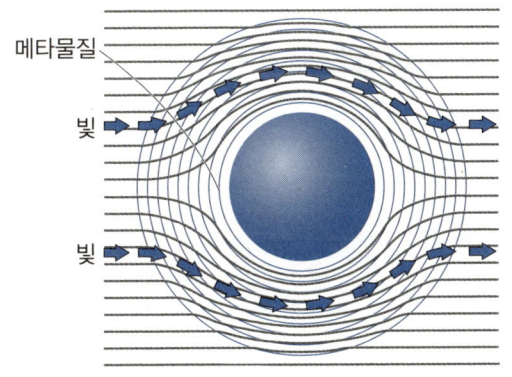

물리군   투명 망토를 씌우면 물체에 빛이 닿지 않으니까 물체를 볼 수 없는 거군요.

정교수   맞아. 빛 대신 소리가 물체에 도달하지 못하게 할 수도 있어.

물리군   어떻게요?

정교수   소리는 음파라는 이름의 파동이야. 음파가 물체에 도달하지 못하게 만드는 물질을 음향 메타물질이라고 불러. 음향 메타물질은 금속판을 이용해서 와플 모양으로 만들었지.

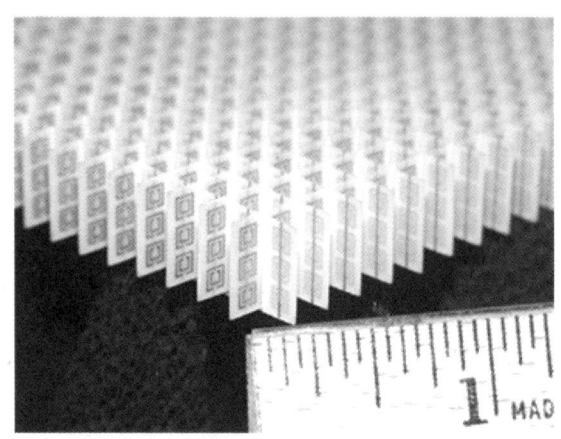

배가 물속에 있는 물체를 확인할 때는 음파(소리)를 사용해. 음파를 물체에 쏜 후 반사된 음파가 배에 도달하는 것을 통해 물속에 어떤 물체가 있다는 걸 알게 되지. 하지만 배에서 쏜 음파가 음향 메타물질로 에워싼 잠수함에 도달하면 반사되지 않고 꺾여 휘어지므로 잠수함을 발견할 수 없어. 이렇게 음향 메타물질로 만든 잠수함을 스텔스 잠수함이라고 부른다네.

반응집광과 양자컴퓨터 _ 양자광학의 활용과 미래

정교수 양자광학은 양자정보과학에서 이용돼. 광자는 양자컴퓨터의 큐비트로 활용되며, 이를 통해 정보를 저장하고 처리하지. 양자컴퓨터는 일반 컴퓨터에 비해 빠르고 효율적인 계산이 가능해.

양자컴퓨터에서는 반응집광을 이용해 광자를 큐비트로 사용한다. 빛의 반응집성은 전송 오류와 보안 위반을 줄여 양자암호화 프로토콜의 보안을 강화한다. 예를 들어 양자 보안 직접 통신(Quantum Secure Direct Communication, QSDC)에서 반응집성을 통해 제공되는 광자의 독특한 시간적 분리는 보안을 강화하여, 도청자가 전송된 정보를 가로채고 디코딩하는 것을 어렵게 만든다.

양자 네트워크의 개발은 먼 거리에 있는 여러 양자 장치를 연결하는 것을 목표로 한다. 이때 정보를 안정적으로 전송하기 위해 단일 광자 소스에 의존한다. 반응집광은 전송된 각 신호가 뚜렷하고 모호하지 않도록 한다. 그리하여 데이터 손실 가능성을 줄이고 양자 네트워크의 전반적인 효율성과 확장성을 향상시킨다. 또한 반응집광은 양자 정보를 전송하고 데이터의 무결성을 보장하는 신뢰할 만한 수단을 제공함으로써 양자 네트워크에서 중요한 역할을 한다.

양자컴퓨터

또한 양자광학은 통신, 보안, 암호화, 센서 등 다양한 분야에 적용된다. 양자 통신은 정보를 안전하게 전송하는 기술로 큰 관심을 받고 있으며, 양자암호화는 보안성이 높은 통신을 가능하게 한다.

반응집광은 측정 감도를 향상시키기 위해 양자 계측에 사용된다. 양자 간섭계는 반응집성의 비고전적 특성을 활용하여 거리, 자기장 및 중력파와 같은 물리량을 보다 정확하게 측정할 수 있다. 이러한 발전으로 천문학, 지구 물리학 및 재료 과학에 널리 쓰인다.

또한 반응집광은 의료 이미징 및 진단에 잠재적으로 응용할 수 있다. 양자 강화 형광 이미징 및 단일 광자 방출 컴퓨터 단층 촬영(SPECT)과 같은 기술은 빛의 반응집성을 활용한다. 이러한 기술에 의해 기존 방법보다 더 높은 해상도와 감도로 질병을 조기에 발견하고 보다 정확한 진단이 가능하다.

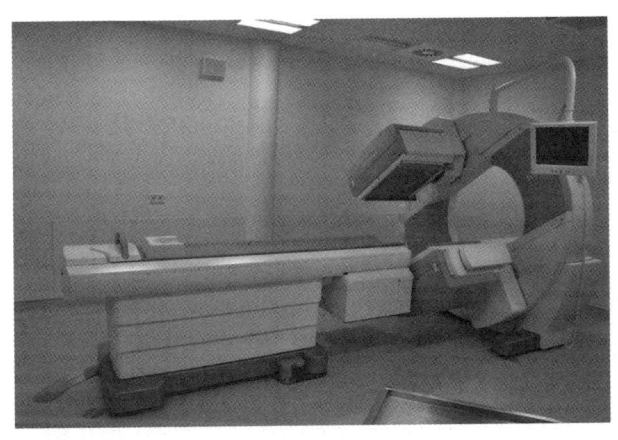

　반응집광은 광학 컴퓨팅에 사용되는 단일 광자 소스를 개발하는 데 매우 중요하다. 광학 컴퓨팅에서 정보는 전기 신호가 아닌 빛을 사용하여 처리되고 전송된다. 따라서 기존 전자 시스템보다 처리 속도가 빠르고 에너지 소비가 낮은 장점이 있다.

　반응집광 기술에 대한 전망은 매우 밝다. 반응집광 소스는 양자 메모리 및 프로세서에 이용할 수 있다. 또한 이 기술은 보안 통신, 고급 컴퓨팅, 정밀 감지 및 기초 물리학 연구의 혁신을 주도할 것으로 예상된다. 반응집광은 양자광학의 중추로 자리매김할 것이다.

만남에 덧붙여

Comparison of Quantum and Semiclassical Radiation Theories with Application to the Beam Maser*

E. T. JAYNES[†] AND F. W. CUMMINGS[‡]

Summary—This paper has two purposes: 1) to clarify the relationship between the quantum theory of radiation, where the electromagnetic field-expansion coefficients satisfy commutation relations, and the semiclassical theory, where the electromagnetic field is considered as a definite function of time rather than as an operator; and 2) to apply some of the results in a study of amplitude and frequency stability in a molecular beam maser.

In 1), it is shown that the semiclassical theory, when extended to take into account both the effect of the field on the molecules and the effect of the molecules on the field, reproduces almost quantitatively the same laws of energy exchange and coherence properties as the quantized field theory, even in the limit of one or a few quanta in the field mode. In particular, the semiclassical theory is shown to lead to a prediction of spontaneous emission, with the same decay rate as given by quantum electrodynamics, described by the Einstein A coefficients.

In 2), the semiclassical theory is applied to the molecular beam maser. Equilibrium amplitude and frequency of oscillation are obtained for an arbitrary velocity distribution of focused molecules, generalizing the results obtained previously by Gordon, Zeiger, and Townes for a singel-velocity beam, and by Lamb and Helmer for a Maxwellian beam. A somewhat surprising result is obtained; which is that the measurable properties of the maser, such as starting current, effective molecular Q, etc., depend mostly on the slowest 5 to 10 per cent of the molecules.

Next we calculate the effect of amplitude and frequency of oscillation, of small systematic perturbations. We obtain a prediction

---

* Received September 28, 1962.
† Washington University, St. Louis, Mo.
‡ Aeronutronic, Division of Ford Motor Co., Newport Beach, Calif.

that stability can be improved by adjusting the system so that the molecules emit all their energy $\hbar\Omega$ to the field, then reabsorb part of it, before leaving the cavity. In general, the most stable operation is obtained when the molecules are in the process of absorbing energy from the radiation as they leave the cavity, most unstable when they are still emitting energy at that time.

Finally, we consider the response of an oscillating maser to randomly time-varying perturbations. Graphs are given showing predicted response to a small superimposed signal of a frequency near the oscillation frequency. The existence of "noise enhancing" and "noise quieting" modes of operation found here is a general property of any oscillating system in which amplitude is limited by nonlinearity.

I. INTRODUCTION

THIS PAPER has two purposes: 1) to clarify the relationship between the quantum theory of radiation where the electromagnetic field expansion coefficients satisfy commutation relations, and the semiclassical theory where the electromagnetic field is considered as a definite function of time rather than as an operator, and 2) to apply some of the results thus obtained in a study of amplitude and frequency stability of the ammonia beam maser.

In 1), the relation between quantum electrodynamics and the semiclassical theory is shown to be quite different from that usually assumed. The semiclassical theory, when extended to take into account both the effect of the molecules on the field and the effect of the field on the molecules, reproduces almost quantitatively the same laws of energy exchange and coherence properties as the quantized field theory, even in the limit of one or a few quanta in the field cavity mode. In particular, the semiclassical theory is shown to lead to a prediction of spontaneous emission, with exactly the same decay rate as given by quantum electrodynamics, as described by the Einstein A coefficients.

There remain, however, several fundamental differences in the two theories. For example, quantum electrodynamics allows the possibility that the combined system (molecules plus field) may be in states which have properties qualitatively different than any that can be described in classical terms, even in the limit of arbitrarily high photon occupation numbers. Thus the common statement that quantum electrodynamics goes over into classical electrodynamics in the case of high quantum numbers for the field oscillators, needs to be somewhat qualified.

Having shown the essential equivalence of quantum electrodynamics and the semiclassical approach for the problems of interest, we turn to detailed calculations applying the semiclassical theory to the ammonia beam maser. Equilibrium amplitude and frequency of oscillation are obtained for an arbitrary velocity distribution of focused molecules, generalizing the results obtained previously by Gordon, Zeiger and Townes [1] for a single-velocity beam and by Lamb and Helmer [6] for a Maxwellian beam. A rather surprising result is obtained, namely that the measurable properties of the maser, such as starting current, effective molecular Q, etc., depend mostly on the slowest 5 to 10 per cent of the molecules.

Next we calculate the effect on amplitude and frequency of oscillation of small systematic perturbations. We obtain a prediction that stability can be improved by adjusting the system so that the molecules emit all their energy $\hbar\Omega$ to the field, then reabsorb part of it, before leaving the cavity. In general, the most stable operation is obtained when the molecules are in the process of absorbing energy from the radiation as they leave the cavity, the most unstable, when they are still emitting energy at that time.

Finally, we consider this response of an oscillating maser to random time-varying perturbations. Graphs are given showing predicted response to a small superimposed signal of a frequency near to the oscillation frequency. The results show a quite complicated variation as a function of the frequency difference and beam current, and resemble some results of Wiener, concerning nonlinear random phenomena.

Broadly speaking, there are two different levels of approximation used in gas maser theories published as of this writing:

1) The most common and also the crudest of these theories is the one wherein one treats the emission process of radiation from molecules as if the transition probabilities were proportional to the time. Such theories contain little that was not already contained in Einstein's 1917 paper which introduced the A and B coefficients. According to quantum mechanics, the idea of time proportional transition probabilities is an approximation, valid only when the correlation time of the radiation is short compared to the time required to accumulate an appreciable transition probability; that is, the radiation responsible for the transition must be random, with a spectrum wide compared to the line width. In an ammonia beam device the correlation time of the radiation may be of the order of 10^3 to 10^6 times the flight time of a molecule through the cavity, and thus any attempt to describe maser operation in terms of "Fermi golden rule" type of equations for the transition probabilities, i.e.,

$$W_{1\to 2} = \frac{2\pi}{\hbar^2} \mid H_{12}\mid^2 \rho(\omega)$$

may lead to conclusions qualitatively as well as quantitatively wrong. Most of the existing noise figure calculations are based on a treatment of this type [1], [2], and hence one cannot assess their worth until the calculation has been checked by a more rigorous theory.

2) The second method of treating the maser theoretically is that based on solving Schrödinger's time-dependent equation for a molecule as perturbed by a classically described field and finding then the expectation value of the dipole moment of the molecule and using the time derivative of this expectation value as the current source of the classical electromagnetic field.

This is essentially the calculation of Shimoda, Wang and Townes [4], and Basov and Prokhorov [5], and Lamb and Helmer [6], Feynman, Vernon and Hellwarth [7]. While this is clearly superior to the first method outline above, there are still several important approximations involved. In principle, the molecular beam should be treated as a single quantum-mechanical system, by a formalism like that of Dicke's "superradiant gas" [8]. In the theories quoted above, the molecules were ascribed independent wave functions. Also, the electromagnetic field should be quantized and the problem treated as one of quantum electrodynamics. Although the theories above lead to definite predictions for saturation and frequency pulling, it is not at all clear that they can lead to reliable predictions of fluctuation effects involved in noise figure and frequency stability. It is generally thought that the semiclassical theory should be adequate for any effects at microwave frequencies due to the smallness of the Einstein A coefficient compared to the B coefficient. However, quantization of the electromagnetic field introduces many changes in addition to the appearance of A coefficients; for instance, quantization can lead to states qualitatively different from any describable in classical terms, even in the limit of arbitrarily high photon occupation numbers per field normal mode. Such states will be shown, in the calculations to follow, to actually be the ones produced in the maser under certain idealized conditions. Thus until these calculations based on these approximations are checked in some other way, our degree of confidence in them cannot be too great.

Our approach in this paper will be, stated briefly, to first treat simple problems in which we can talk of transition probabilities with all coherence properties retained, within the formalism of quantum electrodynamics. Then we will investigate the relationship between the "modified" semiclassical ("neoclassical") theory, as employed by Shimoda, Wang, and Townes [4], and quantum electrodynamics. The relationship is not at all that which is usually assumed, *i.e.*, that quantum electrodynamics goes into semiclassical theory only in the limit of high photon occupation numbers per field-normal mode. Rather the neoclassical theory, in which expectation values of quantum mechanical operators are interpreted as *actual* values of sources in the classical Maxwell equations, and both the effect of the radiation field on the molecule and the effect of the molecule back on the field are taken into account, *does* lead to a prediction of spontaneous emission, and to only very small quantitative differences in the decay rate for the case of a few microwave photons in the cavity.

In Section IV the neoclassical theory is applied to the problem of the ammonia beam device in which the indirect "coupling" between molecules via the field is treated and a steady-state solution is obtained under the assumption of an arbitrary velocity distribution, wall losses and/or external energy coupling. The solution is obtained for the frequency stability as a function of the mean values of the square and cube of the flight time and the Q of the cavity. This solution is found to agree with that of Gordon, Zeiger and Townes [1] in the univelocity case and with the analysis of Lamb and Helmer [6] in the case of a Maxwellian distribution of velocities. Then this solution is made the basis, or unperturbed solution, in a perturbation treatment of fluctuation effects to the first order in the small departure from the steady-state solution. These problems in this case become linear, so that we can analyze the effect of small periodic perturbations proportional to $\exp(i\beta t)$ and superpose the solutions to give solutions for the transient response to an arbitrary small perturbation. This can represent an extra signal fed in intentionally, or it might be a randomly varying function representing thermal noise in the cavity and/or load. A "noise quieting" phenomena is seen to occur for proper values of the flight time, and graphs are drawn which exhibit the power spectrum of the thermal noise as affected by the molecular beam.

II. Quantum Electrodynamic Solutions

We approach the theory of maser operation in several stages, starting with simple, special cases for which all details of the mathematics can be worked out, then adding various features which tend in the direction of more realistic models. The mathematical form of the theory is quite similar to what one encounters in the statistical mechanics of irreversible processes. Of particular interest, however, is the extent to which the semiclassical theory is derivable from quantum electrodynamics, and we are most interested in comparing the results of this section with those obtained in Section III. Also, the effect of different statistical assumptions concerning the initial states of the molecules is interesting in this same regard.

A. *Field Quantization*

We first develop the formalism of field quantization in a form suitable for microwave applications. There is, of course, no need for elegant covariant formulations here; the simple approach to electrodynamics given by Fermi [9] is quite adequate for our purposes. Here the usual plane-wave expansion is not appropriate and in its place we need to use the expansion of electromagnetic fields in terms of resonant modes of the particular cavity under consideration. We use the cavity normal mode functions as defined by Slater [10]. The cavity is represented by a volume V, bounded by a closed surface S. Let $E_a(x)$, $k_a^2 = \omega_a^2/C^2$ be the eigenfunctions and eigenvalues of the boundary-value problem.

$$\nabla \times \nabla \times E - k^2 E = 0 \quad \text{in } V$$
$$n \times E = 0 \quad \text{on } S \quad (1)$$

where n is a unit vector normal to S. The $E_a(x)$ are so normalized that

$$\int_v (E_a \cdot E_b) dV = \delta_{ab}. \tag{2}$$

The vector functions $H_a(x)$, related to E_a by

$$\nabla \times E_a = k_a H_a \qquad \nabla \times H_a = k_a E_a \tag{3}$$

are also orthonormal in V as follows:

$$\int_v (H_a \cdot H_b) dV = \delta_{ab}. \tag{4}$$

The electric and magnetic fields can be expanded in the following forms:

$$E(x\ t) = -\sqrt{4\pi} \sum_a p_a(t) E_a(x) \tag{5}$$

and

$$H(x\ t) = \sqrt{4\pi} \sum_a \omega_a q_a(t) H_a(x). \tag{6}$$

From these relations, we find for the total field energy

$$\mathcal{H} = \int \frac{E^2 + H^2}{8\pi} dV = \frac{1}{2} \sum_a (p_a^2 + \omega_a^2 q_a^2), \tag{7}$$

and the Maxwell equations,

$$\nabla \times E = -\frac{1}{c} \frac{\partial H}{\partial t} \tag{8}$$

and

$$\nabla \times H = \frac{1}{c} \frac{\partial E}{\partial t}, \tag{9}$$

then reduce to the Hamiltonian equations of motion,

$$\dot{q}_a = \frac{\partial \mathcal{H}}{\partial p_a} = p_a, \tag{8a}$$

$$\dot{p}_a = -\frac{\partial \mathcal{H}}{\partial q_a} = -\omega_a^2 q_a, \tag{8b}$$

respectively.

On quantization of the field the canonically conjugate coordinates and momenta satisfy the commutation rules,

$$[q_a, q_b] = [p_a, p_b] = 0 \tag{10}$$

and

$$[q_a, p_b] = i\hbar \delta_{ab}. \tag{11}$$

The operators C_a^*, C_a which create or annihilate a photon in the ath cavity mode are then

$$C_a^* = \frac{p_a + i\omega_a q_a}{\sqrt{2\hbar\omega_a}} \qquad C_a = \frac{p_a - i\omega_a q_a}{\sqrt{2\hbar\omega_a}} \tag{12}$$

with the commutation rule

$$[C_a, C_b^*] = \delta_{ab}. \tag{13}$$

Denote by $\phi(n_1, n_2, \cdots)$ the state vector of the field for which there are n_1 quanta in mode 1, n_2 in mode 2, etc. The C_a operators have the properties

$$C_a \phi(\cdots, n_a, \cdots) = \sqrt{n_a}\, \phi(\cdots, n_a - 1, \cdots) \tag{14}$$

and

$$C_a^* \phi(\cdots, n_a, \cdots)$$
$$= \sqrt{n_a + 1}\, \phi(\cdots, n_a + 1, \cdots) \tag{15}$$

from which we easily verify (13), and obtain the matrix elements in the n_a representation,

$$(n_a | C_a | n_a') = (n_a' | C_a^* | n_a)$$
$$= \sqrt{n_a + 1}\, \delta(n_a'\ n_a + 1). \tag{16}$$

The Hamiltonian, with zero point energy removed, then reduces to

$$\mathcal{H} = \sum_a \hbar \omega_a C_a^* C_a = \sum_a \hbar \omega_a n_a. \tag{17}$$

Finally, we work out for later purposes the matrix elements of the electric field in the case of a cylindrical cavity with only the lowest TM mode excited. In this mode, the only nonvanishing component of E_a is E_{az} = (constant) $\times J_0(k_a r)$, independent of z and θ. The normalizing constant is obtained from evaluating the integral (2), with the result that on the axis of the cylinder (along which the molecules travel in an ammonia maser) the function E_{az} reduces to

$$E_{az} = \frac{1}{J_1 \sqrt{V}}. \tag{18}$$

Here $J_1 = J(u) = 0.5191$, and $u = 2.405$ is the first root of $J_0(u) = 0$. V is the volume of the cavity. The operator p_a involved in the electric field expansion is, from (12),

$$p_a = \sqrt{\frac{\hbar\omega_a}{2}}(C_a + C_a^*). \tag{19}$$

Combining (5), (16), (18), and (19), we obtain the matrix elements

$$(n | E | n')$$
$$= -\left(\frac{2\pi\hbar\omega}{J_1^2 V}\right)^{1/2} [\sqrt{n}\, \delta_{n,n'+1} + \sqrt{n+1}\, \delta_{n+1,n'}] \tag{20}$$

in which we have dropped the subscript a, it being understood that (20) refers to the case where only the lowest TM mode is taken into account. For the matrix elements of electric field at points off the axis of the cylinder, this expression should be multiplied by $J_0(Kr)$.

B. Interaction with a Single Molecule

The simplest possible situation is one where we consider a lossless cavity, which has only a single resonant mode near the natural line frequency of the molecule, and a uniform field (electric or magnetic, whichever is the one effective in field-molecule interaction) along the path of the molecules. Suppose further that only a single molecule, which has only two possible energy levels, is in the cavity. With the molecule-field interaction in the usual ($J \cdot A$) form, it appears that even this problem cannot be solved exactly. However, because of the simplicity of the model, we will be able to treat it more accurately than is usually done in more difficult problems, where one resorts to an expansion in powers of ($e^2/\hbar c$). The stationary states of the system (molecule plus field) can be found to an accuracy of perhaps one part in 10^7 for radiation energy densities up to the order of those encountered in masers, by a calculation which involves nothing worse than solving quadratic equations. By use of perturbation theory still better accuracy would be feasible, but this is not done here.

Let the two possible energy levels of the molecule be denoted by E_m, and the corresponding states by $\psi_m (m=1, 2)$. Similarly, the number of quanta in the field oscillator will be n, and the corresponding state of the field by $\phi (n=0, 1, 2, \cdots)$. The state vectors $\psi_m \phi_n$ then form a basis for the system (molecule plus field). In this representation, the total Hamiltonian is

$(mn \mid H \mid m'n')$
$= (E_m + n\hbar\omega)\delta_{mm'}\delta_{nn'} + (mn \mid H_{\text{int}} \mid m'n').$ (21)

The interaction Hamiltonian between molecule and field is taken of the form

$$H_{\text{int}} = -\mu \cdot E \quad (22)$$

where μ is the electric dipole moment of the molecule, whose component along E shall have the matrix elements

$$(mn \mid \mu_z \mid m'n') = \mu(1 - \delta_{mm'})\delta_{nn'}. \quad (23)$$

Combining this with (20), we obtain the matrix elements for the interaction energy

$(mn \mid H_{\text{int}} \mid m'n')$
$= \hbar\alpha(1 - \delta_{mm'})[\sqrt{n}\,\delta_{n,n'+1} + \sqrt{n+1}\,\delta_{n+1,n'}]$ (24)

where

$$\alpha = \frac{\mu}{J_1}\sqrt{\frac{2\pi\omega}{\hbar}} \quad (25)$$

is the interaction constant. Using the value [11] $\mu = 1.47 \times 10^{-18}$ esu for ammonia, and a cavity 10 cm long, we find $(\alpha/\omega) = 2.08 \times 10^{-10}$ or, $\alpha \approx 5.0$ cps.

The interaction Hamiltonian has matrix elements of two different types: $H_{\text{int}} = V + W$, where

$V_n = (1, n+1 \mid V \mid 2, n) = (2, n \mid V \mid 1, n+1)$
$= \hbar\alpha\sqrt{n+1}$

and

$W_n = (1, n \mid W \mid 2, n+1) = (2, n+1 \mid W \mid 1, n)$
$= \hbar\alpha\sqrt{n+1}$ (26)

all other elements being zero. The term V cannot be treated as a perturbation, for its matrix elements connect "unperturbed" states with an energy separation ($E_2 - E_1 - \hbar\omega$) which goes through zero as the cavity is tuned exactly on the natural line frequency. On the other hand, elements of W connect states with unperturbed energy separation $(E_2 - E_1 + \hbar\omega) \approx 2\hbar\omega$. Since in typical operation conditions ($n \approx 10^6$) we have $W_n 2\hbar\omega < 10^{-7}$, we may treat W as a small perturbation, or even neglect it entirely. We thus write the Hamiltonian as

$$H = H_0 + W$$

in which the term $H_0 = (H_{\text{mol}} + H_{\text{field}} + V)$ must be diagonalized exactly. This is readily done, since H_0 has a "block form" consisting of many (2×2) matrices along the main diagonal. The eigenvalues and eigenfunctions of H_0, defined by $H_0 \Phi_n^\pm = E_n^\pm \Phi_n^\pm$, are the ground state

$$E_0 = E_1 = \hbar\omega_0 \qquad \Phi_0 = \psi_1 \phi_0 \quad (27)$$

and for $n > 0$,

$E_n^\pm = \hbar\omega_n^\pm = \frac{1}{2}[E_1 + E_2 + (2n-1)\hbar\omega]$
$\pm \frac{1}{2}[(E_2 - E_1 - \hbar\omega)^2 + 4n\hbar^2\alpha^2]^{1/2}.$ (28)

We find it convenient now to define our zero molecular energy midway between the levels E_1 and E_2 such that

$$E_1 + E_2 = 0 \qquad E_2 - E_1 = \hbar\Omega$$

so that (28) now reads

$E_n^\pm = \hbar\omega_n^\pm = (n - \frac{1}{2})\hbar\omega \pm \frac{\hbar}{2}[(\Omega - \omega)^2 + 4n\alpha^2]^{1/2}.$ (28a)

Now

$\Phi_n^+ = \psi_2 \phi_{n-1} \cos\theta_n + \psi_1 \phi_n \sin\theta_n$
$\Phi_n^- = -\psi_2 \phi_{n-1} \sin\theta_n + \psi_1 \phi_n \cos\theta_n$ (29)

where

$$\tan 2\theta_n = \frac{2\alpha\sqrt{n}}{\omega - \Omega}. \quad (30)$$

We now require the time-development matrix (in units with $\hbar = 1$)

$$U(t, t') = U(t - t') = \exp[-iH(t - t')] \quad (31)$$

for which the perturbation expansion is

$$U(t) = e^{-iH_0 t} - i\int_0^t e^{-i(t-t')H_0} W e^{it'H_0} dt' + \cdots. \quad (32)$$

The major term $U_0 = \exp(-iH_0t)$ has the matrix elements, for $n > 0$,

$(2, n-1 | U_0 | 2, n-1)$
$\quad = a_n = \cos^2\theta_n e^{-i\omega_n^+ t} + \sin^2\theta_n e^{-i\omega_n^- t}$

$(2, n-1 | U_0 | 1, n)$
$\quad = b_n = \sin\theta_n \cos\theta_n (e^{-i\omega_n^- t} - e^{-i\omega_n^+ t})$

$(2, n | U_0 | 2, n-1) = b_n$

$(1, n | U_0 | 1, n)$
$\quad = c_n = \cos^2\theta_n e^{-i\omega_n^- t} + \sin^2\theta_n e^{-i\omega_n^+ t}$ (33)

and, for $n = 0$,

$(1, 0 | U_0 | 1, 0) = e^{-i\omega_0 t}$ (34)

where now $\omega_0 = -\Omega/2$ all other elements vanish. The transition probability for emission or absorption of one photon during time t is therefore, neglecting terms in W,

$|b_n|^2 = \sin^2 2\theta_n \sin^2(\omega_n^+ - \omega_n^-)t/2 = \dfrac{n\alpha^2 \sin^2 \beta t}{\beta^2}$ (35)

where

$4\beta^2 = (\omega - \Omega)^2 + 4n\alpha^2.$ (36)

The above notation has been chosen in such a way that the block form of U_0 consists of the symmetric, (2×2) unitary matrices

$\begin{bmatrix} a_n & b_n \\ b_n & c_n \end{bmatrix} \quad n = 1, 2, \cdots$

along the main diagonal. The first row and column, however, contain only the single term (34).

We now consider the effect on the field of passing a single molecule through the cavity, with flight time τ. At the instant ($t = 0$) when the molecule enters the cavity, let its state be described by the density matrix $\rho_1(0)$, and the state of the field by the density matrix $\rho_f(0)$. The initial density matrix fo the entire system is thus the direct product $\rho(0) = \rho_1(0) \times \rho_f(0)$, with matrix elements

$(mn | \rho(0) | m'n') = (m | \rho_1(0) | m')(n | \rho_f(0) | n').$ (37)

During the interaction, ρ undergoes a unitary transformation

$\rho(t) = U(t, 0) \rho(0) U^{-1}(t, 0)$ (38)

and the density matrix $\rho_f(t)$, which describes the state of the field only, is the projection[1] of (38) onto the space of the field variables

$(n | \rho_f(t) | n') = \sum_m (mn | \rho(t) | mn').$ (39)

[1] This formalism is developed in detail by Jaynes [12].

The net change in the state of the field thus consists of a linear transformation,

$(n | \rho_f(t) | n') = \sum_{k,k'} (nn' | G | kk')(k | \rho_f(0) | k')$ (40)

or

$\rho_f(\tau) = G \rho_f(0)$ (41)

$(nn' | G | kk')$
$\quad = \sum_{m, m', m''} (m''n | U | mk)(m'k' | U^{-1} | m''n') \sigma_{mm'}$ (42)

where we have written for brevity

$\sigma_{mm'} \equiv (m | \rho_1(0) | m').$ (43)

The sums (42) are readily evaluated with the use of (33), with the result that the only nonvanishing elements of G are

$(nn' | G | n, n') = a_{n+1} a_{n'+1}^* \sigma_{22} + c_n c_n^* \sigma_{11}$ (44a)

$(nn' | G | n+1, n') = b_{n+1} a_{n'+1}^* \sigma_{12}$ (44b)

$(n, n' | G | n, n'+1) = a_{n+1} b_{n'+1}^* \sigma_{21},$ (44c)

$(n, n' | G | n, n'-1) = c_n b_{n'}^* \sigma_{12},$ (44d)

$(n, n' | G | n-1, n') = b_n c_{n'}^* \sigma_{12},$ (44e)

$(n, n' | G | n+1, n'+1) = b_{n+1} b_{n'+1}^* \sigma_{11},$ (44f)

$(n, n' | G | n-1, n'-1) = b_n b_{n'}^* \sigma_{22}.$ (44g)

These relations hold for all quantum numbers n if we understand that c_0 is not defined by (33) but by $c_0 = \exp(-i\omega_0 t)$.

To illustrate the use of this formalism, we discuss a few simple problems using (44). Consider first the case where the field is initially in its lowest state; $(0 | \rho_f(0) | 0) = 1$, all other elements of $\rho_f(0)$ vanish. Then according to (44), after a molecule with initial density matrix σ has passed through, the field density matrix has elements

$(0 | \rho_f(\tau) | 0) = |a_1|^2 \sigma_{22} + \sigma_{11}$

$(0 | \rho_f(\tau) | 1) = (1 | \rho_f(\tau) | 0)^* = c_0 b_1^* \sigma_{12}$

$(1 | \rho_f(\tau) | 1) = |b_1|^2 \sigma_{22},$ (45)

all other elements still vanishing. If the molecule were initially in its lowest state then nothing happens, and the field remains in its ground state. If the molecule was initially in the upper state $[\sigma_{22} = 1, \sigma_{11} = \sigma_{12} = 0]$ we have a simple transition probability of $|b_1|^2$ for the molecule to emit one photon in passing through. If there was initially no coherence relation between upper and lower states of the molecule, then $\sigma_{12} = 0$, and ρ_f remains diagonal; no coherence between states $n = 0$ and $n = 1$ can be set up by the molecule unless there was some coherence initially between upper and lower states of the molecule.

The expectation value of electric field along the axis of the cavity, as obtained from (20), is

$$\langle E \rangle = \text{Trace } (\rho_f E)$$

$$= -\frac{\hbar\alpha}{\mu}\sum_n \sqrt{n+1}[(n|\rho_f|n+1) + (n+1|\rho_f|n)]$$

$$= -\frac{2\alpha\hbar}{\mu}\text{Re}\sum_n \sqrt{n+1}(n|\rho_f|n+1). \quad (46)$$

This remains zero as long as there is no coherence among adjacent levels, even though the energy stored in the field may be large. In the case (45), we obtain for $\langle E \rangle$,

$$\langle E \rangle = -\frac{2\alpha\hbar}{\mu}\text{Re }(c_0 b_1{}^*\sigma_{12})$$

$$= 2\hbar\alpha^2 \frac{\sin\beta t}{\mu\beta}\text{Re }[i\sigma_{12}e^{i(\Omega+\omega)t/2}] \quad (47)$$

where β is defined by (36) with $n=1$. Suppose now the cavity is so tuned that its resonant frequency ω is equal to Ω, then $\beta = \alpha$ and we obtain simply

$$\langle E \rangle = \frac{2\hbar\alpha}{\mu}\sin\alpha t\text{ Re }[i\sigma_{12}e^{i\omega t}]. \quad (47a)$$

Since $\alpha \approx 5$ cps, the term $\sin(\alpha t)$ reaches its first maximum in a quarter cycle, or about 1/20 of a second. This is the interaction time required for a molecule to emit a photon, with probability one, into a lossless cavity initially in its ground state. This shows the great enhancement of spontaneous emission probability due to the presence of the resonant cavity, for the same molecule in empty space would emit with a natural line width (full width at half-maximum intensity),

$$\Delta\omega = \frac{8\omega^3\mu^2}{3\hbar c^3} \approx 10^{-7}\text{ sec}^{-1}, \quad (48)$$

which leads to spontaneous emission times of the order of months at the frequencies here considered.

If the molecule and field are in arbitrary initial states, the general transformation of the field caused by passage of the molecule is, from (44),

$(n|\rho_f(t)|n')$
$= \sigma_{11}[b_{n+1}b_{n'+1}^*(n+1|\rho_f(0)|n'+1) + c_n c_n^*(n|\rho_f(0)|n')]$
$+ \sigma_{12}[b_{n+1}a_{n'+1}^*(n+1|\rho_f(0)|n') + c_n b_{n'}^*(n|\rho_f(0)|n'-1)]$
$+ \sigma_{12}[a_{n+1}b_{n'+1}^*(n|\rho_f(0)|n'+1) + b_n c_{n'}^*(n-1|\rho_f(0)|n')]$
$+ \sigma_{22}[a_{n+1}a_{n'+1}^*(n|\rho_f(0)|n')$
$\quad\quad + b_n b_{n'}^*(n-1|\rho_f(0)|n'-1)]. \quad (49)$

If the field density matrix is initially diagonal,

$$(n|\rho_f(0)|n') = \rho_n \delta_{nn'}. \quad (50)$$

The only nonvanishing components of $\rho_f(t)$ are

$(n|\rho_f(t)|n+1) = \sigma_{11}[|b_{n+1}|^2\rho_{n+1} + |c_n|^2\rho_n]$
$\quad + \sigma_{22}[|a_{n+1}|^2\rho_n + |b_n|^2\rho_{n-1}] \quad (51)$

and

$(n|\rho_f(t)|n+1) = (n+1|\rho_f(t)|n)^*$
$\quad = \sigma_{12}[b_{n+1}a_{n+2}^*\rho_n + c_n b_{n+1}^*\rho_n]. \quad (52)$

These relations will be used in the next section.

C. Successive Single-Molecule Interactions

If several molecules pass through the cavity in succession, the Nth entering as the $(N-1)$th leaves, all with the same initial state, this generates a Markov chain,

$$\rho_f(N\tau) = G^N\rho_f(0) = G\rho_f(N\tau - \tau). \quad (53)$$

Of particular interest is the limit $N \to \infty$.

If the density matrices of field and molecule are initially diagonal,

$$\sigma_{12} = \sigma_{21} = 0 \quad (n|\rho_f(0)|n') = \rho_n\delta_{nn'}, \quad (54)$$

then ρ_f remains diagonal for all time. In this case the entering molecules can always be described by a temperature, defined by

$$\sigma_{22} = \sigma_{11}e^{-x} = (e^x + 1)^{-1}$$
$$x = \hbar\Omega/kT \quad (55)$$

and, using (51), (53) reduces to

$\rho_n(N\tau) = (e^x + 1)^{-1}[(|a_{n+1}|^2 + |c_n|^2 e^x)\rho_n(N\tau - \tau)$
$\quad + |b_{n+1}|^2 e^x\rho_{n+1}(N\tau - \tau) + |b_n|^2\rho_{n-1}(N\tau - \tau)]. \quad (56)$

From this the limiting form of ρ_n may be found. Taking note of the fact that fact that $|a_n|^2 + |b_n|^2 = |b_n|^2 + |c_n|^2 = 1$, we find that a necessary and sufficient condition for a steady state $\rho_n(N\tau) = \rho_n(N\tau - \tau) = \rho_n$, is that the quantities

$$B_n = |b_n|^2(\rho_{n-1} - e^x\rho_n)$$

be independent of n. Now $\Sigma_n\rho_n = 1$, and so $\rho_n \to 0$ as $n \to \infty$. Consequently, $B_n \to 0$, since $|b_n|^2 \leq 1$. Thus B_n can be independent of n only if $B_n = 0$, and the only steady-state solution is the Boltzmann distribution,

$$\rho_n = e^{-x}\rho_{n-1}, \quad (57)$$

for all n for which $|b_n|^2 \neq 0$. From (35) it is seen that b_n could vanish only for isolated special values of n.

Note that (57) is not a Boltzmann distribution with the same temperature T as that of the molecules, except in the case where the cavity is tuned exactly to the natural line frequency. The temperature of (57) is $T_f = \omega T/\Omega$. This difference would never be seen in practice because as soon as one detunes the cavity appreciably the transition probability $|b_n|^2$ becomes extremely small, and the temperature of the radiation would be determined by its interaction with the walls of the cavity, here neglected.

Nevertheless, in principle the difference is there, and we have an example of an interaction between two systems which maintains them at different temperatures. The origin of the phenomena lies in the fact that we have described the state of the molecule in terms of a temperature, which is not wholly justified, since nothing was said about their kinetic energy of translational motion. It is this translational motion which supplies or absorbs the excess energy so as to remove the above apparent violation of energy conservation. When a molecule enters or leaves the cavity it passes through a region of inhomogeneous field, and experiences a net force which very slightly changes its velocity.

In the "negative temperature" case where the entering molecules are more likely to be in the upper state, $\sigma_{22} > \sigma_{11}$, and $x < 0$, the solution $B_n = $ constant is still formally the only stationary one. But it now represents an infinite amount of energy in the field and could never be reached by any finite number of molecules passing through the cavity. It is, of course, only our neglect of losses which leads to such a result, and in practice the operating level quickly reaches a steady value which can be predicted by adding a phenomenological damping term to $\dot{\rho}$ in a well-known way.

As long as the density matrix σ of the entering molecules is diagonal, the density matrix of the field alone also remains diagonal; the expectation value of the electric field remains zero in spite of the fact that the number of photons present may be very large. That is, $\langle E^2 \rangle$ can be very large but $\langle E \rangle$ remains zero. This is more or less to be expected since the entering molecules do not "tell" the field what phase to have. This situation raises certain questions, however, regarding the relation between quantum theory and classical theory. It is usually supposed that the condition for validity of classical electromagnetic theory is simply that the number of photons in each normal mode is large, and that then one may identify the classical electromagnetic field with the quantum-mechanical expectation value. It is seen, however, that this is a necessary but not sufficient condition, for here we have a situation where the semiclassical theory of radiation could not describe such states.

The statement, found in most books on quantum theory, that in the limit of large quantum numbers, quantum theory goes over into classical theory is somewhat misleading. Actually it is possible by coherent superposition of quantum states to construct states which are not describable in terms of classical theory at all. Thus it is that we arrive at the conclusion that classical theory is but a special case of quantum theory in the case of large quantum numbers, *i.e.*, large quantum numbers are necessary but not sufficient to insure the transition from quantum to classical theory.

The case in which two molecules pass through the cavity with flight time τ and leave just as two others enter, etc., has been worked out. The mathematics is tedious, and the result is substantially the same as the successive single-molecule interaction case.

III. RELATION BETWEEN QUANTUM ELECTRODYNAMICS AND SEMICLASSICAL RADIATION THEORY

A. *Semiclassical Electrodynamics*

Now one considers that the electric field $E(t)$ is classically describable, and introduces a wave function,

$$\psi(t) = a(t)\psi_1 + b(t)\psi_2, \qquad (58)$$

for the molecule alone, which develops in time according to the Schrödinger equation

$$i\hbar\dot{\psi} = (H_{\text{mol}} + H_{\text{int}})\psi \qquad (59)$$

where

$$(m|H_{\text{mol}}|m') = E_m \delta_{mm'}. \qquad (60)$$

and

$$(m|H_{\text{int}}|m') = (m| - \mathbf{\mu} \cdot E(t)|m')$$
$$= -\mu(1-\delta_{mm'})E(t). \qquad (61)$$

Schrödinger's equation (59) then reduces to

$$i\hbar\dot{a} = E_1 a - \mu E(t) b$$
$$i\hbar\dot{b} = -\mu E(t) a + E_2 b. \qquad (62)$$

These equations describe the effect of the field on the molecule.

Semiclassical theory as usually treated does not consider the effect of the molecule on the field. To find the effect of the molecule on the field, one calculates the expectation value of the dipole moment of the molecule from the solution of (62),

$$M(t) \equiv \langle \mathbf{\mu} \rangle(t) = \mathbf{\mu}(ab^* + ba^*), \qquad (63)$$

and assumes that the field satisfies the classical equations of motion which would result from interaction with a dipole of moment $M(t)$. This is obtained most easily from the Hamiltonian equations of motion by addition of the interaction energy

$$-M \cdot E = +\sqrt{4\pi} \sum_a p_a(t) E_a(x) \cdot M(t) \qquad (64)$$

to $\mathcal{H}$ in (7) of Section II, where x denotes the position of the molecule. The classical equations of motion are now

$$\dot{p}_a = -\frac{\partial \mathcal{H}}{\partial q_a} = -\omega_a^2 q_a$$

and

$$\dot{q}_a = \frac{\partial \mathcal{H}}{\partial p_a} = p_a + \sqrt{4\pi} M \cdot E_a(x). \qquad (65)$$

Eliminating q_a,

$$\ddot{p}_a + \omega_a^2 p_a = - \sqrt{4\pi}\, \omega_a^2 \mathbf{M} \cdot \mathbf{E}_a(\mathbf{x}). \qquad (66)$$

Assuming that we have only one normal mode excited, the electric field of this mode satisfies the differential equation

$$\ddot{E} + \omega^2 E = \frac{+4\pi\omega^2}{J_1^2 V} M \qquad (67)$$

where again we drop the subscript a. If the cavity has a finite Q, due to wall losses and/or energy coupled out, this is taken into account by adding a phenomenological damping term to (67), giving us

$$\ddot{E} + \frac{\omega}{Q}\dot{E} + \omega^2 E = \frac{4\pi\omega^2 M}{J_1^2 V}. \qquad (68)$$

By the "semiclassical" theory we mean the system of equations (62), (63) and (68). They may be given a somewhat neater formal appearance by eliminating the amplitudes $a(t)$, $b(t)$. The result is the nonlinear system of coupled equations,

$$\ddot{M} + \Omega^2 M = - K^2 W E, \qquad (69a)$$
$$\dot{W} = E\dot{M} \qquad (69b)$$

and

$$\ddot{E} + \omega/Q \dot{E} + \omega^2 E = SM, \qquad (69c)$$

where

$$K = 2\mu/\hbar \qquad S = 4\pi\omega^2/J_1^2 V \qquad (70)$$

and

$$W = E_1 |a|^2 + E_2 |b|^2 - \tfrac{1}{2}(E_1 + E_2)$$
$$= \frac{\hbar\Omega}{2}(|b|^2 - |a|^2) \qquad (71)$$

is the expectation value of energy of the molecule, referred to a zero lying midway between the levels E_1, E_2. In the form (69) we have an apparently classical nonlinear system, all reference to "quantum-mechanical" quantities having disappeared.

The first two equations of (69) admit a first integral,

$$\dot{M}^2 + \Omega^2 M^2 + K^2 W^2 = \text{const.} = \left(\frac{K\hbar\Omega}{2}\right)^2. \qquad (72a)$$

This is readily verified by eliminating E between them. Eq. (72a) is a disguised form of the principle of conservation of probability, $|a|^2 + |b|^2 = 1$. Similarly, the last two equations of (69) can be combined, in the case $Q = \infty$, to yield the constant of the motion

$$\dot{E}^2 + \omega^2 E^2 + 2S(W - ME) = \text{constant}, \qquad (72b)$$

which is easily identified as the conservation of energy statement for the system.

B. The Relation Between Semiclassical and Quantum Electrodynamic Equations of Motion

For the equation of motion of any quantum-mechanical operator we have $i\hbar\dot{F} = [F, H]$. Differentiating this, we have

$$\hbar^2 \ddot{F} + [H, [H, F]] = i\hbar[\dot{H}, F] \qquad (73)$$

which is exact for any operator F which has no explicit time dependence. Let us apply this identity to the electric field operator $F = E$. The total Hamiltonian $H = (H_{\text{mol}} + H_{\text{field}} + H_{\text{int}})$ has no explicit time dependence, so the right-hand side of (73) will vanish. To evaluate the double commutator, we note that H_{int} commutes with E but not with $[H_f, E]$, while H_m commutes with both. Therefore,

$$[H, [H, E]] = [H_f, [H_f, E]] + [H_{\text{int}}, [H_f, E]]. \qquad (74)$$

These commutators are easily worked out, and the result is

$$[H_f, [H_f, E]] = \hbar^2 \omega^2 E \qquad (75)$$
$$[H_{\text{int}}, [H_f, E]] = - \hbar^2 S \mu_{\text{op}}. \qquad (76)$$

Thus a special case of (73) is the operator identity

$$\ddot{E} + \omega^2 E = S\mu_{\text{op}} \qquad (77)$$

which is to be compared to (12c). If we interpret (12c) as the expectation value of (20), they are seen to be identical in the limit $Q \to \infty$, provided that the expectation value of μ_{op} be defined, not in terms of $a(t)$ and $b(t)$ by means of (6), but as the expectation value taken over the complete density matrix $(mn/\rho/m'n')$, i.e.,

$$\langle \mu_{\text{op}} \rangle = \text{Tr}\,(\rho \mu_{\text{op}}) = \sum_{nmm'} (mn\rho m'n)(m' | \mu_{\text{op}} | m). \qquad (78)$$

With this change in interpretation (69c) is seen to be an exact consequence of quantum electrodynamics.

We now write out the identity (73) for the operator $F = \mu_{\text{op}}$. This time H_{int} commutes with μ_{op}, but not with $[H_m, \mu_{\text{op}}]$, while H_f commutes with both. Therefore,

$$[H, [H, \mu_{\text{op}}]] = [H_m, [H_m, \mu_{\text{op}}]] + [H_{\text{int}}, [H_m, \mu_{\text{op}}]]. \qquad (79)$$

Proceeding as before, a short calculation yields the following results:

$$[H_m, [H_m, \mu_{\text{op}}]] = \hbar^2 \Omega^2 \mu_{\text{op}} \qquad (80)$$

and

$$[H_{\text{int}}, [H_m, \mu_{\text{op}}]] = \hbar^2 K^2 H' E \qquad (81)$$

where we have defined an operator

$$H' = H_{\text{mol}} - \tfrac{1}{2}(E_1 + E_2) \qquad (82)$$

with matrix elements

$$(mn | H' | m'n') = \frac{\hbar\Omega}{2}(-1)^m \delta_{mm'} \cdot \delta_{nn'}, \qquad (83)$$

which is just the energy of the molecule, referred to a zero lying midway between its levels E_1, E_2. Combining these relations, we find that another special case of (73) is the operator identity

$$\ddot{\mu}_{op} + \Omega^2 \mu_{op} = - K^2 H' E \qquad (84)$$

which is to be compared to (69a). However, now when we take the expectation value of (84) we do not get (69a) in general, for in the semiclassical equation the "driving term" appears as $\langle H'\rangle\langle E\rangle$, while quantum electrodynamics yields $\langle H'E\rangle$. The difference between these terms arises from the possibility of having correlated states, a situation inherent in quantum electrodynamics but not in semiclassical theory. When the states of field and molecule are uncorrelated, the density matrix reduces to a direct product $\rho = \rho_m \rho_f$, or

$$(mn|\rho|m'n') = (m|\rho_m|m')(n|\rho_f|n') \qquad (85)$$

when (85) holds, then $\langle H'E\rangle = \langle H'\rangle\langle E\rangle$. But in general, $\langle H'E\rangle \neq \langle H'\rangle\langle E\rangle$.

The possibility of obtaining "correlated states" can arise whenever two or more quantum-mechanical systems interact. Quantum electrodynamics allows the possibility of states of the combined system (molecule plus field) which are in a definite pure state, but nevertheless one cannot ascribe any definite quantum state to the molecule alone, or the field alone. This possibility forms the basis of one of Einstein's objections to quantum mechanics. The Einstein-Podolsky-Rosen [13] paradox consists of the fact that when such correlated states exist, one has the possibility of predicting with certainty either one of two noncommuting quantities of a system by making measurements which do not involve any physical interaction with it.

An interesting line of thought is based on the fact that the semiclassical theory and quantum electrodynamics predict different equations of motion for a molecule in the field, the difference arising just from those correlated states which cause the above conceptual difficulties. Thus if one could find any experimental situation in which the difference between $\langle H'E\rangle$ and $\langle H'\rangle\langle E\rangle$ leads to any observable difference in maser operations, this would constitute an indirect, but convincing, check on those aspects of quantum theory which lead to the Einstein-Podolsky-Rosen paradox. However, as will be shown, the prospects of detecting such a difference are extremely dubious, for we will see that the semiclassical theory actually reproduces many of the features which one commonly supposes can be found only with field quantization.

C. Solution of Nonlinear Semiclassical Equations

The simplest approximate solution of the coupled semiclassical equations is the one wherein we ignore the time variation of W, thereby converting the problem into a linear one, similar to the case of two coupled pendulums. The normal modes are found by assuming that E and M have a common time factor $\exp(i\nu t)$; if $W=$ constant, then (69a) and (69c) reduce to

$$(\omega^2 - \nu^2)(\Omega^2 - \nu^2) + K^2 SW = 0 \qquad (86)$$

or

$$\nu^2 = \frac{\omega^2 + \Omega^2}{2} \pm \frac{1}{2}\sqrt{(\omega^2 - \Omega^2)^2 - 4K^2 SW}. \qquad (87)$$

We see here a new feature, not present in coupled pendulums. If $W > 0$ and the cavity is tuned so closely to the natural line frequency that

$$|\omega^2 - \Omega^2| < \sqrt{4K^2 SW}, \qquad (88)$$

the square root in (87) becomes imaginary; one of the normal modes grows exponentially, the other decays. Now an oscillation of growing amplitude represents energy being transferred from molecule to field, and therefore we see that the semiclassical theory *does* lead to a prediction of spontaneous emission. Since W is just the energy of the molecule, we see that the condition of unstable growing oscillation is just that the molecule's wave function contains more of the upper state than the lower, $|b|^2 > |a|^2$.

Suppose that the cavity is tuned exactly to the natural line frequency, $\omega = \Omega$. Then (87) reduces to

$$\nu^2 = \omega^2 \pm i\sqrt{K^2 SW} \qquad (89)$$

or to an extremely good approximation,

$$\nu = \omega \pm \frac{i\sqrt{K^2 SW}}{2\omega}. \qquad (90)$$

If we start with the molecule nearly in the upper state then $W = \hbar\Omega/2$ and the amplitude of the field varies like

$$\exp\left(\frac{\sqrt{K^2 SW}}{2\omega}t\right)e^{i\omega t} = \exp \alpha t e^{i\omega t} \qquad (91)$$

where α is the interaction constant defined in (25). This is to be compared to the result (47a) describing spontaneous emission according to quantum electrodynamics. It is seen that although the two approaches lead to equations of different functional form, they predict exactly the same characteristic time $1/\alpha$ for spontaneous emission. This shows that the relation between quantum electrodynamics and the semiclassical theory of radiation is quite different from what is usually supposed. Physically, it means that whenever the molecule has a dipole moment different from zero, the fields set up by this dipole react back on the molecule and change its state in such a way that energy is delivered to the field, as long as $W > 0$. These linear relations do not hold indefinitely, of course. From the conservation law (72a) it is clear that when the amplitude of the M

oscillation increases, the magnitude of W must decrease, and this will eventually put a stop to the emission process.

The change in time of the variable W, for any reasonable value of field strength, as we are concerned with here, is slow compared to the time variation of E or M, and for typical ammonia maser operating conditions, E and M go through the order of 10^7 cycles for each cycle of W. For a qualitative picture of the slow changes in the case $\omega = \Omega$, we may consider the orbits in the $(\dot{E}, \omega E)$ plane and in the $(\dot{M}, \omega M)$ plane, as in Fig. 1. Noting that the interaction energy is typically about 10^{-6} times smaller than the energy of the molecule W, the conservation of energy law (72b) reduces, in almost all cases, to

$$(\dot{E})^2 + \omega^2 E^2 + 2SW = \text{constant}, \qquad (92)$$

which shows that as W increases, the orbit in the $(\dot{E}, \omega E)$ plane must shrink, and vice versa. Also, the conservation law (72a) shows that if $|W|$ increases, the M orbit must shrink, and vice versa. Therefore the direction of all secular changes is determined by the sign of W and $\dot{W}$. In the equation $\dot{W} = E\dot{M}$ we can for all practical purposes replace $E\dot{M}$ by its average over one cycle, $\overline{E\dot{M}}$, since we are interested in the trend of W over time scales of many cycles, rather than small rapid fluctuations whose effect averages to zero over a cycle. Secular changes in W depend, thus, only on the sign of $\overline{E\dot{M}}$.

Whenever the E motion is advanced in phase over the M motion, we have $\overline{E\dot{M}} \geq 0$. In this case, W will slowly increase and the E orbit will shrink. The M orbit will then grow if $W<0$, shrink if $W>0$. If the M motion is advanced in phase over the E motion, all these changes are reversed. The situation is summarized by the orbit diagrams of Fig. 2. Or again, let us assume that W is given as some periodic function of time, so that we can summarize these same conclusions graphically as in Fig. 3.

Whenever the E orbit is expanding, energy is being delivered from the molecule to the field, and the necessary and sufficient condition for this is that the M motion be advanced in phase over the E motion. Thus in order to understand the long time course of events, one must study the secular changes in relative phases of the E and M motion.

To this end introduce the slowly varying complex amplitudes X and Y, defined by

$$\dot{E} + i\omega E = X(t)e^{i\omega t} \qquad (93)$$

and

$$\dot{M} + i\omega M = Y(t)e^{i\omega t}. \qquad (94)$$

The quantities depicted in Fig. 2 are just the complex numbers (93) and (94). Noting the properties,

$$(\dot{E})^2 + \omega^2 E^2 = |X|^2 \qquad (95)$$

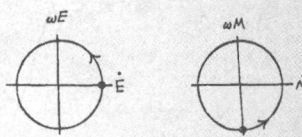

Fig. 1—Closed orbits in the phase space of the E and M oscillators. The dots indicate that the E motion is 90° ahead of the M motion in the phase.

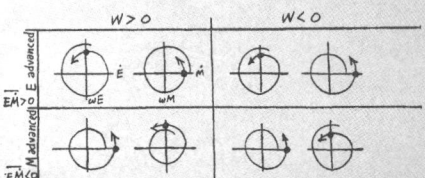

Fig. 2—Secular changes in orbits for the four combinations of signs of $\overline{E\dot{M}}$ and W.

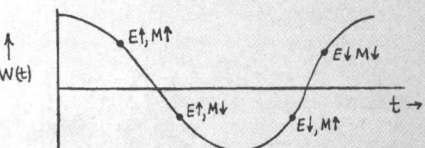

Fig. 3—Representation of the orbit diagrams in another way.

and

$$\dot{E} + \omega^2 E = \dot{X}e^{i\omega t} \qquad (96)$$

and similarly for M, we can write the equations of motion (69) in the form, for the case $\omega = \Omega$,

$$2i\omega \dot{X} = S(Y - Y^* e^{-2i\omega t}), \qquad (97a)$$

$$2i\omega \dot{Y} = -K^2 W(X - X^* e^{-2i\omega t}), \qquad (97b)$$

$$4i\omega \dot{W} = XYe^{2i\omega t} + XY^* - X^*Y - X^*Y^* e^{-2i\omega t}. \qquad (97c)$$

The conservation laws become

$$|Y|^2 + K^2 W^2 = \text{constant} = \left(\frac{Kh\Omega}{2}\right)^2 \qquad (98a)$$

and

$$|X|^2 + 2SW = \text{constant}. \qquad (98b)$$

Now the functions X and Y are slowly varying functions of time, and again it is their average change over many cycles, rather than the very small rapid fluctuations at frequency 2ω, which are of interest. Thus the oscillating terms in (97) can be dropped, since their average over a cycle is negligible compared to their dc components.

The system of equations determining secular changes of both amplitude and phase is, therefore,

$$2i\omega \dot{X} = SY, \quad (99a)$$

$$2i\omega \dot{Y} = -K^2WX, \quad (99b)$$

$$4i\omega \dot{W} = XY^* - X^*Y. \quad (99c)$$

It is easily verified that the conservation laws (98) are exact consequences of (99). Differentiating (99c) once more and making use of the conservation laws, we can eliminate X and Y, obtaining the equation

$$4\omega^2 \dot{W} - 3SK^2W^2 + K^2CW + SK^2\left(\frac{\hbar\Omega}{2}\right)^2 = 0 \quad (100)$$

where C is the constant of the motion (98b). A first integral of (100) can be obtained immediately by multiplication with $\dot{W}$ and integrating

$$2\omega^2(\dot{W})^2 - SK^2W^3 + \frac{K^2CW^2}{2} + SK^2\left(\frac{\hbar\Omega}{2}\right)^2 W$$

$$= \text{constant}. \quad (101)$$

This equation has the form of the Hamilton-Jacobi equation for motion of a particle in a particular potential well. For any motion in which either of the points $W = \pm(\hbar\Omega/2)$ is accessible, we have the constant on the right-hand side of (101) equal to

$$\frac{K^2C}{2}\left(\frac{\hbar\Omega}{2}\right)^2.$$

This is easily seen from (98a), for if $W = \pm(\hbar\Omega/2)$, then $Y = 0$ and $\dot{W} = 0$. For any such motion the cubic polynomial in (101) factors. To see this most easily, introduce the change of variable $(\hbar\Omega/2)z = W$. Then (101) takes the form,

$$b\dot{z}^2 - (z^3 - z - az^2 + a)$$

$$= b\dot{z}^2 - [(z-1)(z+1)(z-a)] = 0, \quad (102)$$

where

$$b = \frac{4\omega^2}{SK^2\hbar\omega} = \frac{1}{2\alpha^2} \quad a = \frac{C}{S\hbar\omega}. \quad (103)$$

The solution is

$$\sqrt{2}\,\alpha t = \int_{z(0)}^{z(t)} \frac{dz}{\sqrt{(1-z)(1+z)(a-z)}}. \quad (104)$$

The z motion is therefore periodic between turning points represented by singularities of the integrand. If $a \gtrsim 1$, these turning points are at $z = \pm 1$, while if $a < 1$ they are at $z = -1$ and $z = a$. Now we consider evaluation of the constant $a = C/\hbar\omega S$. From (5), (9a) and (17) we have

$$\dot{E}^2 + \omega^2 E^2 = \frac{4\pi\omega^2}{J_1^2 V}(2n\hbar\omega)$$

where n is the number of photons stored in the cavity.

Now, examination of (72b) with the small interaction term neglected gives us

$$\frac{4\pi\omega^2}{J_1^2 V}(2n\hbar\omega) + S\hbar\omega = C = S\hbar\omega(2n+1)$$

if we assume there are n molecules in the field when the molecule is in its upper state, $W = +(\hbar\omega/2)$. Thus $a = (2n+1)$. There is in this theory of course no restriction on n to be an integer. The smallest value which a can attain is represented by zero energy in the field and the molecule in its ground state, $a = -1$, or $n = -1$. When n is negative this of course means that the total energy is insufficient for the molecule to get into its upper state, and this is the physical reason why the turning point of the z motion then occurs at $z = a$. The integral (104) is one of the standard forms defining elliptic functions. Using the standard notation sn (u, k), the solution for the case $n \geq 0$ is

$$z(t) = -1 + 2sn^2\left(\sqrt{n+1}\,\alpha t + Q, \frac{1}{\sqrt{n+1}}\right) \quad (105)$$

where

$$Q \equiv sn^{-1}\left(\sqrt{\frac{z(0)+1}{2}}, \frac{1}{\sqrt{n+1}}\right) \quad (106)$$

is the initial phase of the motion. In the limit of large n, the elliptic functions approach trigonometric functions, as is seen most easily from (104). If $a \gg 1$, then (104) reduces to

$$\sqrt{2}\,\alpha t \approx \frac{1}{\sqrt{a}}\int \frac{dz}{\sqrt{1-z^2}} = \frac{1}{\sqrt{a}}\sin^{-1} z(t) + \text{constant},$$

or

$$z(t) \approx \sin(2\sqrt{n}\,\alpha t + \theta). \quad (107)$$

The case $a = 1$, $n = 0$ is a special one, for the integrand of (104) then develops a first-order pole at $z = 1$. The solution (105) is still valid but is no longer periodic; in fact sn $(u, 1)$ is equal to tanh u which approaches ± 1 asymptotically as $u \to \pm \infty$. This represents a case where the energy in the field exactly disappears just as the molecule gets into its upper state, and the final stages of the solution then represent the "shrinking normal mode" of (90), where E is 90° ahead of M. (This phase relation is in fact maintained throughout the part of the motion (105) in which z increases. Throughout the decreasing part, E is 90° behind M.)

The point $W = \hbar\Omega/2$, $z = 1$ is a metastable point of the orbit in this case, for if we start out with exactly the initial comditions $z = 1$, $E = M = 0$ then nothing happens. All time derivatives remain zero and the molecule does not emit. However if there is the slightest change in this initial condition, the growing normal mode of (91) will be started up (unless the phase relations between M and E happen to be just the value for the pure shrinking mode), and eventually the energy of the

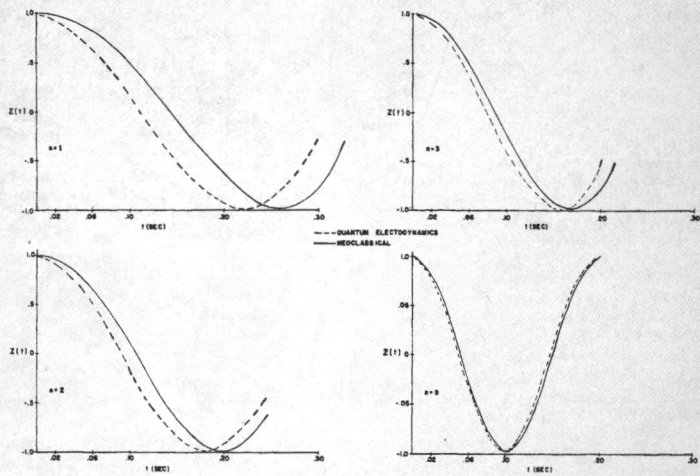

Fig. 4—Energy as a function of time for several n's.

molecule spills out entirely into the field when we reach the lower turning point $z=-1$. The molecule then reabsorbs the energy $\hbar\omega$ from the field, passing back to the metastable point $z=1$ but requiring an infinite time to do so. Fig. 4 shows $z(t)$ as a function of time for several values of the parameter n, the number of photons in the cavity, and these are compared with the corresponding quantum electrodyanmics curves. It is seen that for a few photons, the correspondence is almost exact. Even in the case of one or two quanta, the semiclassical theory gives solutions reproducing almost quantitatively everything that is found in the quantum electrodynamics analysis. Even the "quantum jumps" are still with us, but here they show up as perfectly continuous processes, where an instability develops in the solution of the nonlinear equations and an amount of energy $\hbar\omega$ is more or less rapidly transferred from molecule to field.

The semiclassical analysis gives a very interesting description of the process of spontaneous emission. Consider a large number of molecules, as nearly as possible in the upper state. In practice, of course, we cannot prepare them *exactly* in the upper state, but there will be a certain probability distribution of initial values of amplitude for the growing normal mode. A molecule with an initial value $M_+(0)$ will at time t have an M amplitude of $M_+(0)e^{\alpha t} = M_+(t)$.

If we agree to say that when this reaches the value K, the molecule is actively emitting energy, then, no matter what the probability distribution of initial values, provided only that this distribution is a continuous function in the neighborhood of the metastable point $W = \hbar\Omega/2$, we find that the number of molecules emitting at time t is proportional to exp $(-2\alpha t)$. We can see this by a simple argument which runs as follows: We shall say that the molecule has reached stage K of the emission process when the amplitude of the oscillation reaches a value in the range, $K < M_+(0)e^{\alpha t} < (K + \delta K)$. We now ask, "How many atoms will be in stage K at time t?" Clearly, all those for which $M_+(0)$ lies in the range, $Ke^{-\alpha t} < M_+(0) < (K+\delta K)e^{-\alpha t}$. If the initial probability distribution in the phase space of the molecule is constant, this would be proportional to the area of the annular ring, $2\pi r dr = 2\pi(Ke^{-\alpha t})(\delta Ke^{-\alpha t}) \sim e^{-2\alpha t}$. Thus the "law of radioactive decay" or "time proportional transition probabilities" appears in this analysis as a consequence of the existence of metastable states. The time constant of the decay law is independent of the method of preparation of the molecules, and only depends on the interaction constant of the molecules with the electromagnetic field. The situation is exactly like that of a large number of pencils nearly perfectly balanced on their points. The time required for any one pencil to fall over depends on how close it was to vertical at $t=0$. If the probability distribution of initial states is continuous in the neighborhood of this metastable point, then we have a decay law with a time constant which depends only on the laws of mechanics, not on the method of preparation of initial states.

Mathematically, this semiclassical theory as expounded in this section is exactly the same as that already used by Shimoda, Wang and Townes [4]. The new feature is the realization that this formalism accounts for effects which all standard textbooks describe as requiring field quantization for their explanation. Because of this success, and the fact that the correspondence with quantum electrodynamics continues to strengthen as this formalism is applied to a larger group of problems, it is felt that this formalism deserves independent status as a physical theory in its own right, and we suggest it be called the neoclassical theory of electrodynamics.

Conceptually, the neoclassical theory amounts to reinterpreting the quantities usually denoted as expectation values of energy and dipole moment as *actual* values, the latter serving as the source of classical electromagnetic fields. These fields are then inserted in the Hamiltonian of the molecule and the reaction of the molecule to the field calculated according to the Schrödinger equation. Thus a general problem would be governed by the set of coupled equations, $i\hbar\dot\psi = H(A_u)\psi$ and $\Box A_u + 4\pi\langle j_u\rangle = 0$, where $\langle j_u\rangle$ is the current density operator, the expectation value here being interpreted as a classical current density.

Having now convinced ourselves of the efficacy of this method, we now turn to the application of these equations to the problem of the ammonia beam maser.

IV. Application of Neoclassical Radiation Theory to the Ammonia Maser

A. Ideal Steady-State Solution

Our starting point for obtaining the ideal steady-state solution for the ammonia maser will be (69), *i.e.*,

$$\ddot M_i + \Omega^2 M_i = - K^2 W_i E(t) \quad (108a)$$

$$\ddot E + \omega^2 E + \frac{\omega}{Q}\dot E = SM_i \quad (108b)$$

$$\dot W_i = E\dot M_i \quad (108c)$$

where now the subscript i refers to the ith molecule. If now all of the molecules are subjected to the same field $E(t)$ as in the Stanford [6] ammonia maser, we can simply define the total moment and energy

$$M(t) \equiv \sum_i M_i(t), \quad (109a)$$

$$W(t) \equiv \sum_i W_i(t). \quad (109b)$$

We see that (108a)–(108c) are still satisfied by these quantitites, in particular,

$$\ddot E + \omega^2 E + \frac{\omega}{Q}\dot E = SM = S\sum_i M_i. \quad (110)$$

The conservation law (72a) is still valid in the sense that the left-hand side is still constant; but, of course, the value of the constant now depends on the initial conditions. Since the problem of N molecules in the cavity is hardly any more difficult in this formalism than that of one molecule, we can see the advantage of this formalism over the quantum electrodynamics approach, where two molecules in the cavity at a time would involve solving cubic equations, that of three molecules, quartic equations, and so forth.

As our first application, we consider the case where molecules enter the cavity all in the upper state, $W_i = \hbar\Omega/2$, with the same velocity and with a uniform rate of A molecules per unit of time. We wish to find how, under these circumstances, the steady-state frequency and amplitude of oscillation depend on the experimentally controllable parameters A, ω, Q.

Denote by t_i the time at which the ith molecule enters the cavity. It is readily verified by substitution that the solution of (108a) with the initial conditions $M_i(t_i) = \dot M_i(t_i) = 0$ is

$$M_i(t) = - \frac{K^2}{\Omega}\int_{t_i}^t W_i(t')E(t')\sin\Omega(t - t')dt'. \quad (111)$$

Using this, (108c) can be written as an integral equation. A time integration yields

$$W_i(t) - W_i(t_i) = \int_{t_i}^t dt'' E(t'')\dot M(t'')$$

$$= - K^2\int_{t_i}^t dt'' E(t'')\int_{t_i}^{t''} dt' W_i(t')E(t')\cos\Omega(t'' - t'). \quad (112)$$

Interchanging the order of integration in (112), we find that $W_i(t)$ satisfies an integral equation of Volterra form,

$$W_i(t) - W_i(t_i) = \int_{t_i}^t G(t, t')W_i(t')dt', \quad (113)$$

with the kernel,

$$G(t, t') = - K^2\int_{t'}^t dt'' E(t'')\cos\Omega(t'' - t')E(t'). \quad (114)$$

We now assume the electric field is given by

$$E(t) = 2a\sin\nu t \quad (115)$$

where a and ν are parameters to be determined by the condition that (110), (111) and (113) be self-consistent. It is clear from (108c) that the exact solution of $W_i(t)$ contains terms oscillating at frequencies of the order of $(\Omega+\nu)$. While these terms may contribute appreciably to $\dot W_i$, their effect on W_i averages to zero in times of the order of one cycle of the RF. Since we are interested in the long time drift in W_i, rather than these small rapid fluctuations, we neglect terms in (114) of frequency $(\Omega+\nu)$. Their contribution to W_i is of the relative order of magnitude

$$\frac{(\Omega - \nu)}{(\Omega + \nu)} \lesssim 10^{-7}$$

in all cases of practical interest. With this approximation (114) reduces to

$$G(t, t') \approx - K^2 a^2 \frac{\sin (\Omega - \nu)(t - t')}{(\Omega - \nu)} \quad (116)$$

and the slowly varying part of $W_i(t)$ satisfies the integral equation

$$W_i(t) = W_i(t_i) - K^2 a^2 \int_{t_i}^{t} W_i(t') \frac{\sin (\Omega - \nu)(t - t')}{(\Omega - \nu)} dt'. \quad (117)$$

The exact solution of (117) with the initial condition $W_i(t_i) = \hbar\Omega/2$ is

$$W_i(t) = \frac{\hbar\Omega}{2\lambda^2} [(\Omega - \nu)^2 + a^2 K^2 \cos \lambda(t - t_i)] \quad (118)$$

where

$$\lambda^2 \equiv (\Omega - \nu)^2 + (Ka)^2. \quad (119)$$

As a check, and an illustration of some of our previous remarks, we note that (118) and (119) agree with the results found from quantum electrodynamics, and (35) and (36), and also with the result of others who have treated the problem by direct integration of Schrödinger's equation.

The total dipole moment of all the moledules in the cavity is

$$M(t) = A \int_{t-\tau}^{t} M_i(t) dt_i$$

$$= -\frac{AK^2}{\Omega} \int_{t-\tau}^{t} dt' E(t') \sin \Omega(t - t')$$

$$\cdot \int_{t-\tau}^{t'} dt_i W_i(t') \quad (120)$$

where we have used (111) and inverted the order of integration. With the solution (118) for $W_i(t)$, this becomes

$$M(t) = -\frac{Aa\hbar\Omega K^2}{\Omega\lambda^2} \int_0^{\tau} \left[(\Omega - \nu)^2 q + \frac{a^2 K^2}{\lambda} \sin \lambda q \right]$$

$$\cdot \sin \nu(q + t - \tau) \sin \Omega(\tau - q) x dq \quad (121)$$

where $q = (\tau - t + t')$. As a function of q, the last factor of the integrand contains oscillating terms of frequencies $(\Omega \pm \nu)$, and again the relative contribution of the high-frequency term will be of the order of 10^{-7} or smaller under all conditions of interest. Neglecting this small term, (121) reduces to

$$M(t) = \frac{Aa\hbar\Omega K^2}{\Omega\lambda^2} \Bigg[(1 - \cos \lambda\tau) \cos \nu t$$

$$- \frac{(\Omega - \nu)}{\lambda} (\lambda\tau - \sin \lambda\tau) \sin \nu t \Bigg]. \quad (122)$$

Details from this point on will be considered later, as a special case of a more general solution.

B. Velocity Distribution

In the preceding section, all the molecules were assumed to have the same flight time τ. To find the effect of a velocity distribution, we have only to note that the analysis leading to (122) is still valid, and it gives the contribution to total moment of those molecules with flight time in the range $d\tau$, provided that we replace A by $nf(\tau)d\tau$, where n is the total number of molecules entering the cavity per unit time, and $f(\tau)d\tau$ is the fraction of entering molecules with flight times in the range $d\tau$, normalized so that

$$\int_0^{\infty} f(\tau) d\tau = 1. \quad (123)$$

The total dipole moment of all molecules in the cavity is then obtained by one more integration of (122), as follows:

$$M(t) = n\gamma \Bigg\{ [1 - c(\lambda)] \cos \nu t$$

$$- \frac{(\Omega - \nu)}{\lambda} [\lambda\bar{\tau} - S(\lambda)] \sin \nu t \Bigg\} \quad (124)$$

where we have defined

$$\gamma = \frac{a\hbar\Omega K^2}{\Omega\lambda^2} \quad (125)$$

for convenience, and where

$$\bar{\tau} = \int_0^{\infty} \tau f(\tau) d\tau \quad (126)$$

is the mean flight time, and

$$c(\lambda) \equiv \int_0^{\infty} \cos \lambda\tau f(\tau) d\tau \quad (127)$$

and

$$s(\lambda) \equiv \int_0^{\infty} \sin \lambda\tau f(\tau) d\tau \quad (128)$$

are the Fourier transforms of the flight time distribution.

To obtain the conditions for a self-consistent solution, we substitute (124) into (110) and equate the coefficients of $\cos \nu t$, $\sin \nu t$. We obtain the relations

$$\frac{\omega \nu}{Q} = \frac{Sn\gamma}{2a} [1 - c(\lambda)] \quad (129)$$

and

$$\nu^2 - \omega^2 = \frac{Sn\gamma}{2a} \frac{\Omega - \nu}{\lambda} [\lambda\bar{\tau} - s(\lambda)]. \quad (130)$$

The starting current n_0 is determined by (129) for small λ. From (127) we have

$$\lim_{\lambda \to 0} \frac{1 - c(\lambda)}{\lambda^2} = \int_0^\infty \frac{1}{2} \tau^2 f(\tau) d\tau = \frac{\overline{\tau^2}}{2} \quad (131)$$

so that it is the mean-square flight time which determines the starting current, as follows:

$$n_0 = \frac{4a\omega\nu}{Q\lambda^2 \overline{j\tau^2} S} \approx \frac{\hbar J_1^2 V}{Q\overline{\tau^2} 4\pi\mu^2}. \quad (132)$$

Similarly, we have from (128),

$$\lim_{\lambda \to 0} \frac{\lambda\bar{\tau} - S(\lambda)}{\lambda^3} = \frac{1}{6} \overline{\tau^3}, \quad (133)$$

so that if we define new functions,

$$F(\lambda) \equiv \frac{6[\lambda\bar{\tau} - s(\lambda)]}{\lambda^3 \overline{\tau^3}} \quad (134)$$

and

$$G(\lambda) \equiv \frac{2[1 - c(\lambda)]}{\lambda^2 \overline{\tau^2}}, \quad (135)$$

we have $F(0) = G(0) = 1$. Previous writers [1] have expressed their results in terms of an "effective Q" of the molecular beam. The appropriate definition here would be

$$Q_m \equiv \frac{\Omega \overline{\tau^3}}{6\overline{\tau^2}} \approx 10^6. \quad (136)$$

Our conditions (129) and (130) then assume the forms

$$n_0/n = G(\lambda) \quad (137)$$

and

$$(\nu - \Omega) \cong (\omega - \Omega) \frac{Q}{Q_m} \frac{G(\lambda)}{F(\lambda)} \quad (138)$$

if we neglect terms of order $(Q/Q_m)^2$. These relations are to be used graphically as follows: For a given velocity distribution, the functions $F(\lambda)$ and $G(\lambda)$ can be calculated and plotted. Then from (137) one determines λ as a function of the beam current. The frequency pulling factor (G/F) of (138) is then determined. Finally, the amplitude of oscillation a is determined for given beam current n and cavity tuning ω, by use of (138) and (119). For constant beam current, i.e., constant λ, the graph of amplitude vs frequency of oscillation is a half ellipse, the amplitude reaching a maximum for perfect tuning of the cavity to the molecule line, in which case $Ka = \lambda$. Oscillations can persist over a frequency band $(\Omega - \lambda) < \nu < (\Omega + \lambda)$.

In the case of a single molecular velocity, $f(\tau) = \delta(\tau - \tau_0)$, these relations reduce to those of Shimoda, Wang, and Townes [4], and in the case of a Maxwellian velocity distribution, where the fraction of molecules per unit time with velocities in the range dv is proportional to $v^3 \exp(-v^2/v_0^2) dv$, or

$$f(\tau) \sim \exp(-L^2/v_0^2 \tau^2)/\tau^5 \quad (139)$$

(L = length of cavity, $v_0^2 = 2KT/m$), they reduce to the theory of Lamb and Helmer [6]. In the Maxwellian case our dimensionless parameters F, G, Q_m become identical with those defined by Lamb and Helmer.

The assumption of a Maxwellian distribution is certainly a reasonable first approximation, but it is probably not very accurate because the quality of focusing is velocity dependent, the focuser having the property of focusing the small velocity molecules much more effectively than the high-speed ones; and, except for extremely strong focusing voltage, the distribution of flight times may be biased considerably more in favor of large τ than is indicated by (139). This could have an important effect on stability, as may be seen in the following.

In the case of a Maxwellian distribution, half of the mean-square flight time $\overline{\tau^2}$, which determines the starting current, is contributed by the slowest 12 per cent of the molecules. Half of the third moment $\overline{\tau^3}$, which determines the effective molecular Q, and hence the long time frequency stability, is due to the slowest 1.9 per cent. Any further biasing in favor of higher τ would have a considerable effect on $\overline{\tau^2}$, and a very large effect on $\overline{\tau^3}$. For this reason, effects of fluctuations in beam current may not be of the relative order of magnitude $1/\sqrt{N}$, where N is the total number of molecules in the cavity. Fluctuations in the experimentally significant quantities may be determined almost entirely by the slowest 5 per cent of the molecules, with corresponding greater relative variation.

The effect on the frequency-pulling function G/F of a "truncated" Maxwellian distribution has been worked out for the case where the velocity distribution is taken to be Maxwellian up to some $v_{\max}$ and zero thereafter, where $v_{\max}$ was taken arbitrarily to be $\frac{1}{2}v_0$. The results were compared to that of a Maxwellian distribution. A region of stability still appears, as in the analysis of Lamb and Helmer, but it occurs for smaller values of λ (about a factor of two) which corresponds to smaller values of beam flux.

V. Fluctuation Effects

The steady-state relations found in Section IV form the starting point for investigations of fluctuation effects, by perturbation methods in which we expand in powers of the small departure from the previous solutions. If we calculate only to the lowest nonvanishing order, these problems become linear. But in this case we can analyze the effect of small periodic perturbations, proportional to $\exp(i\beta t)$ and superpose the solutions to find the transient response to an arbitrary small perturbation. Thus, consider the effect of an additional signal $F(t)$ impressed on the cavity. This might be an extra signal intentionally fed in, or it might be a randomly varying function representing thermal noise generated in the cavity and/or load. The equation of motion for the electric field then becomes

$$\ddot{E} + \omega^2 E + \frac{\omega}{Q}\dot{E} = SM(t) + F(t) \qquad (140)$$

and this $F(t)$ causes a change of E_1 in the electric field. Suppose now that $F(t)$ contains the time factor exp $(i\beta t)$; then if β is not too close to the oscillation frequency ν, the change in electric moment of the ith molecule will satisfy

$$\ddot{M}_{1_i} + \Omega^2 M_{1_i} = -K^2 W_i E_1 \qquad (141)$$

where we have set

$$E = E_0 + E_1 = 2a \sin \nu t + E_1 \qquad (142)$$

and

$$M_i = M_{0_i} + M_{1_i} \qquad (143)$$
$$W_i = W_{0_i} + W_{1_i} \qquad (144)$$

where the subscript "0" denotes the unperturbed or steady-state solutions of Section IV. Here we have dropped a term EW_{1_i} on the grounds that it will not have an appreciable component of frequency β. Under these conditions, the change in total moment of all molecules in the cavity will be simply

$$M_1 = \sum_i M_{1_i} = \frac{-K^2 \overline{W} E_1}{\Omega^2 - \beta^2} \qquad (145)$$

where $\overline{W}$ is the average energy of all molecules in the cavity. Combining these relations, we find the electric field fluctuation to be given by

$$E_1 = \frac{F(t)(\Omega^2 - \beta^2)}{\left[\left(\omega^2 - \beta^2 + \frac{i\omega\beta}{Q}\right)(\Omega^2 - \beta^2) + K^2 S \overline{W}\right]} \qquad (146)$$

If $\overline{W}=0$, this reduces, as it must, to the response of the cavity alone. The effect of the molecules is, in this approximation, to suppress the magnitude of the electric field fluctuations E_1, for those frequency components which lie close to the natural line frequency.

If the period of the beat frequency $(\beta-\omega)$ is comparable to the flight time, however, then one should take into account the term W_{1_i}; generally, perturbations of any type can lead to the greatest effects when their frequency is related in this way to the flight time. Since the theory remains linear, fluctuations due to any cause are readily calculated. We now proceed to the calculation of electric field fluctuation in which we retain the term W_{1_i}.

Keeping terms to only the first order in the perturbation, we have to solve the system of linear equations with time varying coefficients,

$$\ddot{E}_1 + \frac{\omega}{Q}\dot{E}_1 + \omega^2 E_1 = SM_1(t) + F(t)$$
$$= S \sum_i M_{1_i}(t) + F(t), \qquad (147)$$

$$\ddot{M}_{1_i} + \Omega^2 M_{1_i} = -K^2 W_{1_i} E_0 - K^2 W_{0_i} E_1, \qquad (148)$$

and

$$\dot{W}_{1_i} = E_1 \dot{M}_{0_i} + E_0 \dot{M}_{1_i}. \qquad (149)$$

From (149) we have

$$W_{1_i}(t) = \int_{t_i}^{t} E_0(t')\left[-K^2 \int_{t_i}^{t'}(W_{1_i}(t'')E_0(t'') + W_{0_i}(t'')E_1(t'')) \cos \Omega(t'-t'')dt''\right]dt'$$
$$+ \int_{t_i}^{t} E_1(t')\left[-K^2 \int_{t_i}^{t'} W_{0_i}(t'')E_0(t'') \cos \Omega(t'-t'')dt''\right]dt' \qquad (150)$$

where we have used

$$M_{1_i}(t) = -\frac{K^2}{\Omega}\int_{t_i}^{t} \sin \Omega(t-t')$$
$$\cdot [W_{1_i}(t')E_0(t') + W_{0_i}(t')E_1(t')]dt'. \qquad (151)$$

We specialize to the case of the tuned cavity, $\omega=\Omega$, since this will not significantly affect the results and renders the mathematics very much less tedious. Now, assuming a solution of the form

$$E_1(t) = a_1 e^{i\beta t} + a_2 e^{i(\beta-2\Omega)t} \qquad (152)$$

we are able to get self-consistent solutions for (147)–(149). Using (152) and the unperturbed solutions of the preceding chapter for the case of a single molecular flight time, we have for (150),

$$W_{1_i}(t) = \frac{Kh\Omega}{4\Omega'}(a_1 - a_2)(e^{-i\Omega' t} - e^{-i\Omega' t_i}) \sin \lambda(t-t_i). \qquad (153)$$

We have dropped terms of frequency $\Omega+\beta$, as they do not contribute appreciably to W_1. Here $\Omega'\equiv\Omega-\beta$. Now we put this result into (151) and after integrating over all the molecules, i.e.,

$$M_1(t) = A\int_{t-\tau}^{t} M_{1_i}(t)dt_i \qquad (154)$$

and, inverting the order of integration, we have

$$M_1(t) = -\frac{K^2 A}{\Omega}\int_{t-\tau}^{t} dt' E_0(t') \sin \Omega(t-t') \int_{t-\tau}^{t'} W_{1i}(t')dt_i$$

$$-\frac{K^2 A}{\Omega}\int_{t-\tau}^{t} dt' E_1(t') \sin \Omega(t-t')$$

$$\cdot \int_{t-\tau}^{t'} W_{0i}(t')dt_i. \tag{155}$$

Note that $\Omega_1 \approx -\Omega_2$ if $\Omega - \beta \leq \pm 10\lambda$. The expressions (157) and (158) can be simplified somewhat by remembering that we can replace A by the starting current A_0 divided by $G(\lambda)$ for the univelocity case, i.e.,

$$A = A_0/G(\lambda) = \frac{4\Omega^3}{Q\tau^2(\hbar\Omega K^2 S)} \frac{\lambda^2\tau^2}{2(1-\cos\lambda\tau)} \tag{163}$$

from (132) and (135). Thus,

$$C = \frac{i\lambda^2(\Omega^2/Q)[\Omega'\cos\Omega'\tau/2 \sin\lambda\tau/2 - \lambda\cos\lambda\tau/2 \sin\Omega'\tau/2]}{2\Omega'(\Omega'^2-\lambda^2)\sin\lambda\tau/2} \tag{164}$$

and

$$B = C + \frac{i\lambda^2\Omega^2/Q[(\cos\lambda\tau-\cos\Omega'\tau) + i/\lambda(\Omega'\sin\lambda\tau - \lambda\sin\Omega'\tau)]}{2\Omega'(\Omega'^2-\lambda^2)(1-\cos\lambda\tau)}. \tag{165}$$

The result of the integrations is

$$SM_1(t) = a_1 Be^{i\beta t} + a_1 Ce^{i(\beta-2\Omega)t} - a_2 Be^{i(\beta-2\Omega)t} - a_2 Ce^{i\beta t} \tag{156}$$

where we have defined

$$C = \frac{AK^2 Shi}{4(\Omega'^2-\lambda^2)} e^{i\Omega'\tau/2}$$

$$\cdot \left[\cos\Omega'\tau/2(1-\cos\lambda\tau) - \frac{\lambda}{\Omega'}\sin\Omega'\tau/2 \sin\lambda\tau\right] \tag{157}$$

and

$$B = C + \frac{AK^2 Shi}{4(\Omega'^2-\lambda^2)}$$

$$\cdot \left[\cos\lambda\tau + i\frac{\Omega'}{\lambda}\sin\lambda\tau - \cos\Omega'\tau - i\sin\Omega'\tau\right]. \tag{158}$$

Inserting this expression into (147) we find

$$a_1 = \frac{F(\Omega_2 + B)}{[(\Omega_1 - B)(\Omega_2 + B) + C^2]} \tag{159}$$

and

$$a_2 = \frac{FC}{[(\Omega_1 - B)(\Omega_2 + B) + C^2]} \tag{160}$$

where we define

$$\Omega_1 = \Omega^2 - \beta^2 + i\Omega\beta/Q \tag{161}$$

and

$$\Omega_2 = \Omega^2 - (\beta-2\Omega)^2 + i\frac{\Omega}{Q}(\beta-2\Omega). \tag{162}$$

We see here the result of having time-varying coefficients in our linear equations (147)–(149). If we feed in a signal of frequency β, the field given back contains the frequencies β and $(2\Omega-\beta)$, i.e., the frequency β and the reflection of β about Ω.

If we plot the amplitudes $|a_1|^2$ and $|a_2|^2$ of the frequencies β and $(2\Omega-\beta)$ as a function of the parameter k, such that $\Omega' = k\lambda$, for various values of the flight time τ, we observe a "quieting" effect for certain values of the flight time. For small values of the flight time, the effect of the molecular beam is to amplify any input signal, e.g., thermal noise. The region of greatest stability, that at which "noise quieting" would be the greatest, appears at $\lambda\tau = 3\pi/2$. See Fig. 5, pp. 107–108.

This analysis was carried out under the assumption of a uniform velocity distribution and a tuned cavity. It is clear enough that these restrictions do not greatly impair the generality of the results obtained. The analysis for an untuned cavity would merely have the effect of shifting the axis in the plots of Fig. 5 by the amount $\Omega-\nu$, so that one would have plots symmetrical about ν instead of Ω. Also, very plausible speculation leads one to conclude that the only effect of a velocity distribution, which mathematically would appear as integrals over τ in B and C in the denominators of (159) and (160), would be to smooth out the "wiggles" which appear near the "bare" cavity response value $FQ/\Omega^2 = 1$.

One can recast the electric field,

$$E(t) = 2a\sin\Omega t + \int_{-\infty}^{+\infty} a_1 e^{i\beta t}d\beta + \int_{-\infty}^{+\infty} a_2 e^{i(2\Omega-\beta)t}d\beta + cc \tag{166}$$

in the form

$$E(t) = A(t)\sin[\Omega t + \phi(t)] \tag{167}$$

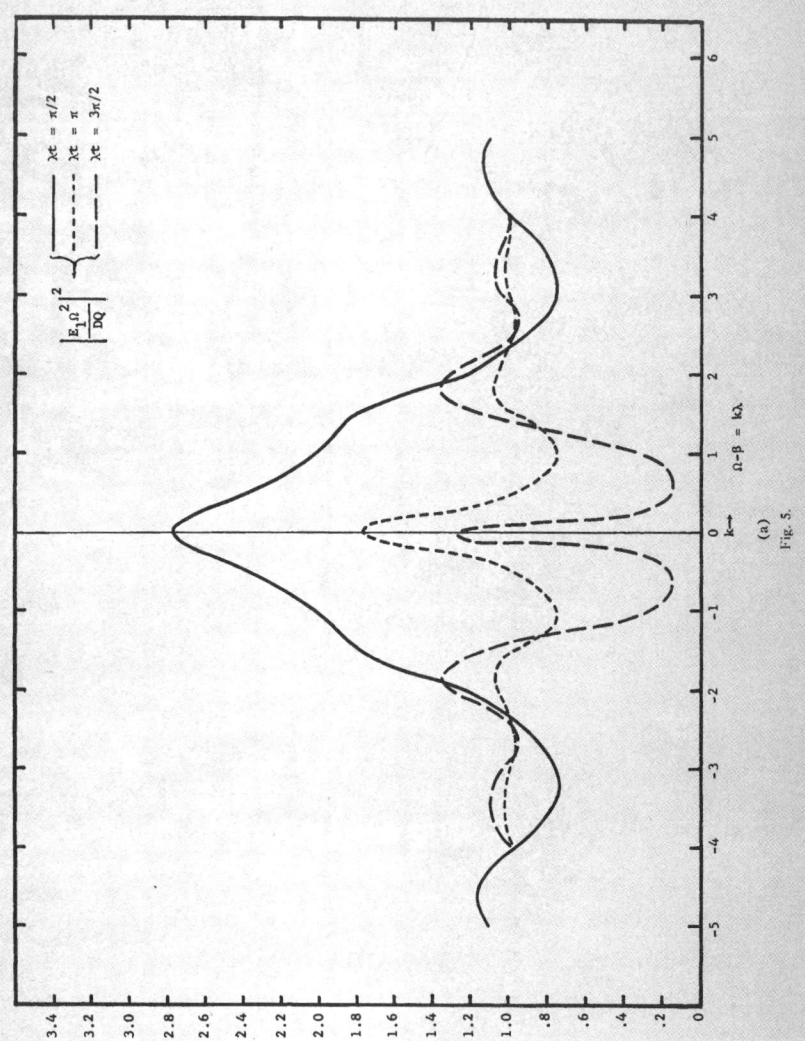

Fig. 5.

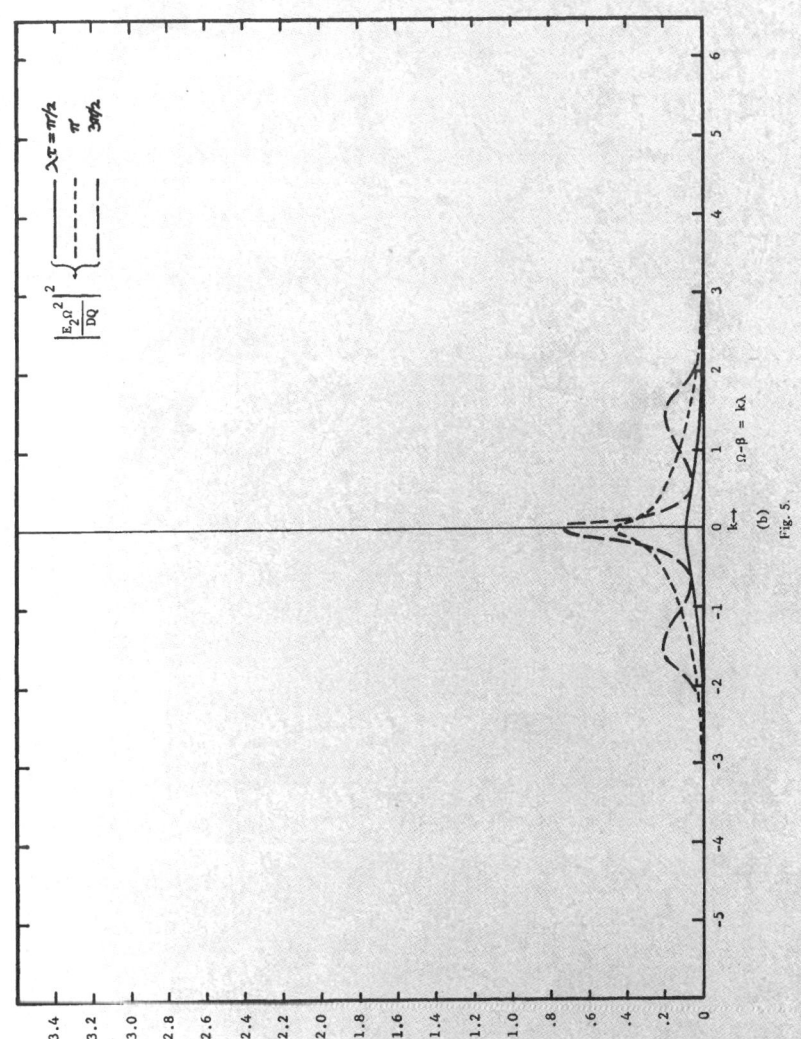

Fig. 5.

where the power spectra of $A(t)$ and $\phi(t)$ are known from the knowledge of a_1 and a_2. One might then argue that we can obtain $\langle \phi^2 \rangle$ and thus have an answer for the frequency stability; but we must remember that we have used a first-order perturbation method, which is valid only for small momentary departures from the steady-state solution—the frequency instability due to thermal noise in the walls probably causes the phase to wander in a manner closely akin to Brownian motion phenomena; and we cannot expect to predict cumulative phenomena by use of perturbations. However, since there is "restoring force" on the amplitude, it can never wander very far from the steady-state value, and we would expect that this analysis could lead to a good prediction of the amplitude stability due to a random perturbation such as thermal noise generated in the cavity or walls.

The criterion for the accuracy of any frequency standard, or "clock," is not so much how far the oscillation frequency drifts from its nominal value in a given time, but rather how well we are able to *predict* how it will wander. Here one would be concerned with the wander of the phase of the output under the perturbation of 1) the thermal radiation from the cavity walls and 2) the unavoidable random fluctuations in beam composition. The analysis given above indicates that under conditions likely to be realized in the foreseeable future, the former effect will be by far the most important.

By analogy with the classical Einstein treatment of Brownian motion, one expects that the phase ϕ will be uncertain in the following sense: While $\langle \phi(t) \rangle = 0$, $\langle \phi^2(t) \rangle = Kt$. The constant K would then be a reasonable measure of the oscillation stability.

The analysis given above is not, however, adapted to answering questions of this type. In assuming that the actual output could be represented in the form

$$E(t) = 2a \sin \nu t + E_1(t)$$

with $|E_1| \ll a$, we have in effect restricted the theory to cases (or time intervals) such that the phase wandering ϕ is at most of order E_1/a. A more general theory of random processes is therefore needed before questions about very long time behavior can be answered.

Wiener [15] has given an interesting discussion of the response of nonlinear systems to a random perturbation. His curves are very similar to those of Fig. 5, which arise, for instance, in the analysis of the alpha rhythm of brain waves. It appears that the phenomena predicted by Fig. 5 is a general property of any nonlinear system which is attempting to stabilize itself.

References

[1] J. P. Gordon, H. J. Zeiger, and C. H. Townes, "Molecular microwave oscillator and new hyperfine structure in the microwave spectrum of NH_3," *Phys. Rev.*, vol. 95, pp. 282–284; July 1, 1954.
——, "The maser—new type of microwave amplifier, frequency standard, and spectrometer," *Phys. Rev.*, vol. 99, pp. 1264–1274; August 15, 1955.
[2] M. W. Muller, "Noise in a molecular amplifier," *Phys. Rev.*, vol. 106, pp. 8–12; April 1, 1957.
[3] J. P. Gordon and L. D. White, "Noise in maser amplifiers—theory and experiment," *Proc. IRE*, vol. 46, pp. 1599–1594; September, 1958.
[4] K. Shimoda, T. C. Wang, and C. H. Townes, "Further aspects of maser theory," *Phys. Rev.*, vol. 102, pp. 1308–1321; June 1, 1956.
[5] N. G. Basov and A. M. Prokhorov, "Applications of molecular beams to radio spectroscopic studies of rotation spectra of molecules," *J. Exp. Theoret. Phys.* (*USSR*), vol. 27, pp. 431–438; 1954. (In Russian.)
——, "On possible methods of producing active molecules for a molecular generator," *J. Exp. Theoret. Phys.* (*USSR*), vol. 28, (Letter), p. 249; 1955.
[6] J. C. Helmer, "Maser oscillators," *J. Appl. Phys.*, vol. 28, pp. 212–215; February, 1957.
[7] R. P. Feynman, F. L. Vernon, Jr., and R. W. Hellwarth, "Geometrical representation of the Schrödinger equation for solving maser problems," *J. Appl. Phys.*, vol. 28, pp. 49–52; January, 1957.
[8] R. H. Dicke, "Coherence in spontaneous radiation processes," *Phys. Rev.*, vol. 93, pp. 99–110; January, 1954.
[9] E. Fermi, "Quantum theory of radiation," *Rev. Mod. Phys.*, vol. 4, pp. 87–102; January, 1932.
[10] J. C. Slater, "Microwave Electronics," D. van Nostrand Co., Inc., New York, N. Y., ch. 4; 1950.
[11] D. K. Coles, et al., "Stark effect of the ammonia inversion spectrum," *Phys. Rev.*, vol. 82, pp. 877–879; June 15, 1951.
[12] E. T. Jaynes, "Information theory and statistical mechanics," *Phys. Rev.*, vol. 108, pp. 171–190; October, 15, 1957.
[13] A. Einstein, B. Podolsky, and N. Rosen, *Phys. Rev.*, vol. 47, p. 777; 1935.
[14] N. Wiener, "Extrapolation, Interpolation and Smoothing of Stationary Time Series," John Wiley and Sons, Inc., New York, N. Y.; 1950.
[15] ——, "Nonlinear Problems in Random Theory," John Wiley and Sons, Inc., New York, N. Y.; 1958.

논문 웹페이지

Coherent and Incoherent States of the Radiation Field*

Roy J. Glauber

Lyman Laboratory of Physics, Harvard University, Cambridge, Massachusetts

(Received 29 April 1963)

Methods are developed for discussing the photon statistics of arbitrary radiation fields in fully quantum-mechanical terms. In order to keep the classical limit of quantum electrodynamics plainly in view, extensive use is made of the coherent states of the field. These states, which reduce the field correlation functions to factorized forms, are shown to offer a convenient basis for the description of fields of all types. Although they are not orthogonal to one another, the coherent states form a complete set. It is shown that any quantum state of the field may be expanded in terms of them in a unique way. Expansions are also developed for arbitrary operators in terms of products of the coherent state vectors. These expansions are discussed as a general method of representing the density operator for the field. A particular form is exhibited for the density operator which makes it possible to carry out many quantum-mechanical calculations by methods resembling those of classical theory. This representation permits clear insights into the essential distinction between the quantum and classical descriptions of the field. It leads, in addition, to a simple formulation of a superposition law for photon fields. Detailed discussions are given of the incoherent fields which are generated by superposing the outputs of many stationary sources. These fields are all shown to have intimately related properties, some of which have been known for the particular case of blackbody radiation.

I. INTRODUCTION

FEW problems of physics have received more attention in the past than those posed by the dual wave-particle properties of light. The story of the solution of these problems is a familiar one. It has culminated in the development of a remarkably versatile quantum theory of the electromagnetic field. Yet, for reasons which are partly mathematical and partly, perhaps, the accident of history, very little of the insight of quantum electrodynamics has been brought to bear on the problems of optics. The statistical properties of photon beams, for example, have been discussed to date almost exclusively in classical or semiclassical terms. Such discussions may indeed be informative, but they inevitably leave open serious questions of self-consistency, and risk overlooking quantum phenomena which have no classical analogs. The wave-particle duality, which should be central to any correct treatment of photon statistics, does not survive the transition to the classical limit. The need for a more consistent theory has led us to begin the development of a fully quantum-mechanical approach to the problems of photon statistics. We have quoted several of the results of this work in a recent note,[1] and shall devote much of the present paper to explaining the background of the material reported there.

Most of the mathematical development of quantum electrodynamics to date has been carried out through the use of a particular set of quantum states for the field. These are the stationary states of the non-interacting field, which corresponds to the presence of a precisely defined number of photons. The need to use these states has seemed almost axiomatic inasmuch as nearly all quantum electrodynamical calculations have been carried out by means of perturbation theory. It is characteristic of electrodynamical perturbation theory that in each successive order of approximation it describes processes which either increase or decrease the number of photons present by one. Calculations performed by such methods have only rarely been able to deal with more than a few photons at a time. The

---

* Supported in part by the U. S. Air Force Office of Scientific Research under Contract No. AF 49(638)-589.

[1] R. J. Glauber, Phys. Rev. Letters **10**, 84 (1963).

description of the light beams which occur in optics, on the other hand, may require that we deal with states in which the number of photons present is large and intrinsically uncertain. It has long been clear that the use of the usual set of photon states as a basis offers at best only an awkward way of approaching such problems.

We have found that the use of a rather different set of states, one which arises in a natural way in the discussion of correlation and coherence[2,3] properties of fields, offers much more penetrating insights into the role played by photons in the description of light beams. These states, which we have called coherent ones, are of a type that has long been used to illustrate the time-dependent behavior of harmonic oscillators. Since they lack the convenient property of forming an orthogonal set, very little attention has been paid them as a set of basis states for the description of fields. We shall show that these states, though not orthogonal, do form a complete set and that any state of the field may be represented simply and uniquely in terms of them. By suitably extending the methods used to express arbitrary states in terms of the coherent states, we may express arbitrary operators in terms of products of the corresponding state vectors. It is particularly convenient to express the density operator for the field in an expansion of this type. Such expansions have the property that whenever the field possesses a classical limit, they render that limit evident while at the same time preserving an intrinsically quantum-mechanical description of the field.

The earlier sections of the paper are devoted to a detailed introduction of the coherent states and a survey of some of their properties. We then undertake in Secs. IV and V the expansion of arbitrary states and operators in terms of the coherent states. Section VI is devoted to a discussion of the particular properties of density operators and the way these properties are represented in the new scheme. The application of the formalism to physical problems is begun in Sec. VII, where we introduce a particular form for the density operator which seems especially suited to the treatment of radiation by macroscopic sources. This form for the density operator leads to a particularly simple way of describing the superposition of radiation fields. A form of the density operator which corresponds to a very commonly occurring form of incoherence is then discussed in Sec. VIII and shown to be closely related to the density operator for blackbody radiation. In Sec. IX the results established earlier for the treatment of single modes of the radiation field are generalized to treat the entire field. The photon fields generated by arbitrary distributions of classical currents are shown to have an especially simple description in terms of coherent states. Finally, in Sec. X the methods of the preceding sections

[2] R. J. Glauber, in Proceedings of the Third International Conference on Quantum Electronics, Paris, France, 1963 (to be published).
[3] R. J. Glauber, Phys. Rev. **130**, 2529 (1963).

are illustrated in a discussion of certain forms of coherent and incoherent fields and of their spectra and correlation functions.

II. FIELD-THEORETICAL BACKGROUND

We have, in an earlier paper,[3] discussed the separation of the electric field operator $\mathbf{E}(\mathbf{r}t)$ into its positive-frequency part $\mathbf{E}^{(+)}(\mathbf{r}t)$ and its negative-frequency part $\mathbf{E}^{(-)}(\mathbf{r}t)$. These individual fields were then used to define a succession of correlation functions $G^{(n)}$, the simplest of which takes the form

$$G_{\mu\nu}^{(1)}(\mathbf{r}t,\mathbf{r}'t') = \mathrm{tr}\{\rho E_\mu^{(-)}(\mathbf{r}t) E_\nu^{(+)}(\mathbf{r}'t')\}, \quad (2.1)$$

where ρ is the density operator which describes the field and the symbol tr stands for the trace. We noted, in discussing these functions, that there exist quantum-mechanical states which are eigenstates of the positive- and negative-frequency parts of the fields in the senses indicated by the relations

$$E_\mu^{(+)}(\mathbf{r}t)|\ \rangle = \mathcal{E}_\mu(\mathbf{r}t)|\ \rangle, \quad (2.2)$$

$$\langle\ |E_\mu^{(-)}(\mathbf{r}t) = \mathcal{E}_\mu^*(\mathbf{r}t)\langle\ |, \quad (2.3)$$

in which the function $\mathcal{E}_\mu(\mathbf{r}t)$ plays the role of an eigenvalue. It is possible, as we shall note, to find eigenstates $|\ \rangle$ which correspond to arbitrary choices of the eigenvalue function $\mathcal{E}_\mu(\mathbf{r}t)$, provided they obey the Maxwell equations satisfied by the field operator $E_\mu(\mathbf{r}t)$ and contain only positive frequency terms in their Fourier resolutions.

The importance of the eigenstates defined by Eqs. (2.2) and (2.3) is indicated by the fact that they cause the correlation functions to factorize. If the field is in an eigenstate of this type we have $\rho = |\ \rangle\langle\ |$, and the first-order correlation function therefore reduces to

$$G_{\mu\nu}^{(1)}(\mathbf{r}t,\mathbf{r}'t') = \mathcal{E}_\mu^*(\mathbf{r}t)\mathcal{E}_\nu(\mathbf{r}'t'). \quad (2.4)$$

An analogous separation into a product of $2n$ factors takes place in the nth- order correlation function. The existence of such factorized forms for the correlation functions is the condition we have used to define fully coherent fields. The eigenstates $|\ \rangle$, which we have therefore called the coherent states, have many properties which it will be interesting to study in detail. For this purpose, it will be useful to introduce some of the more directly related elements of quantum electrodynamics.

The electric and magnetic field operators $\mathbf{E}(\mathbf{r}t)$ and $\mathbf{B}(\mathbf{r}t)$ may be derived from the operator $\mathbf{A}(\mathbf{r}t)$, which represents the vector potential, via the relations

$$\mathbf{E} = -\frac{1}{c}\frac{\partial \mathbf{A}}{\partial t}, \quad \mathbf{B} = \nabla \times \mathbf{A}. \quad (2.5)$$

We shall find it convenient, in discussing the quantum states of the field, to describe the field by means of a discrete succession of dynamical variables rather than a continuum of them. For this reason we assume that the field we are discussing is confined within a spatial volume of finite size, and expand the vector potential within that volume in an appropriate set of vector mode functions. The amplitudes associated with these oscillation modes then form a discrete set of variables whose dynamical behavior is easily discussed.

The most convenient choice of a set of mode functions, $\mathbf{u}_k(\mathbf{r})$, is usually determined by physical considerations which have little direct bearing on our present work. In particular, we need not specify the nature of the boundary conditions for the volume under study; they may be either the periodic boundary conditions which lead to traveling wave modes, or the conditions appropriate to reflecting surfaces which lead to standing waves. If the volume contains no refracting materials, the mode function $\mathbf{u}_k(\mathbf{r})$, which corresponds to frequency ω_k, may be taken to satisfy the wave equation

$$\nabla^2 \mathbf{u}_k + \frac{\omega_k^2}{c^2} \mathbf{u}_k = 0 \qquad (2.6)$$

at interior points. More generally, whatever the form of the wave equation or the boundary conditions may be, we shall assume that the mode functions form a complete set which satisfies the orthonormality condition

$$\int \mathbf{u}_k^*(\mathbf{r}) \cdot \mathbf{u}_l(\mathbf{r}) d\mathbf{r} = \delta_{kl}, \qquad (2.7)$$

and the transversality condition

$$\nabla \cdot \mathbf{u}_k(\mathbf{r}) = 0. \qquad (2.8)$$

The plane-wave mode functions appropriate to a cubical volume of side L may be written as

$$\mathbf{u}_k(\mathbf{r}) = L^{-3/2} \hat{e}^{(\lambda)} \exp(i\mathbf{k} \cdot \mathbf{r}), \qquad (2.9)$$

where $\hat{e}^{(\lambda)}$ is a unit polarization vector. This example illustrates the way in which the mode index k may represent an abbreviation for several discrete variables, i.e., in this case the polarization index ($\lambda = 1, 2$) and the three Cartesian components of the propagation vector $\mathbf{k}$. The polarization vector $\hat{e}^{(\lambda)}$ is required to be perpendicular to $\mathbf{k}$ by the condition (2.8), and the permissible values of $\mathbf{k}$ are determined in a familiar way by means of periodic boundary conditions.

The expansion we shall use for the vector potential takes the form

$$\mathbf{A}(\mathbf{r}t) = c \sum_k \left(\frac{\hbar}{2\omega_k}\right)^{1/2}$$
$$\times (a_k \mathbf{u}_k(\mathbf{r}) e^{-i\omega_k t} + a_k^\dagger \mathbf{u}_k^*(\mathbf{r}) e^{i\omega_k t}), \qquad (2.10)$$

in which the normalization factors have been chosen to render dimensionless the pair of complex-conjugate amplitudes a_k and $a_k^\dagger$. In the classical form of electromagnetic theory these Fourier amplitudes are complex numbers which may be chosen arbitrarily but remain constant in time when no charges or currents are present. In quantum electrodynamics, on the other hand, these amplitudes must be regarded as mutually adjoint operators. The amplitude operators, as we have defined them, will likewise remain constant when no field sources are active in the system studied.

The dynamical behavior of the field amplitudes is governed by the electromagnetic Hamiltonian which, in rationalized units, takes the form

$$H = \tfrac{1}{2} \int (\mathbf{E}^2 + \mathbf{B}^2) d\mathbf{r}. \qquad (2.11)$$

With the use of Eqs. (2.7,8) and of a suitable set of boundary conditions on the mode functions, the Hamiltonian may be reduced to the form

$$H = \tfrac{1}{2} \sum_k \hbar \omega_k (a_k^\dagger a_k + a_k a_k^\dagger). \qquad (2.12)$$

This expression is the source of a well-known and extremely fruitful analogy between the mode amplitudes of the field and the coordinates of an assembly of one-dimensional harmonic oscillators. The quantum mechanical properties of the amplitude operators a_k and $a_k^\dagger$ may be described completely by adopting for them the commutation relations familiar from the example of independent harmonic oscillators:

$$[a_k, a_{k'}] = [a_k^\dagger, a_{k'}^\dagger] = 0, \qquad (2.13a)$$
$$[a_k, a_{k'}^\dagger] = \delta_{kk'}. \qquad (2.13b)$$

Having thus separated the dynamical variables of the different modes, we are now free to discuss the quantum states of the modes independently of one another. Our knowledge of the state of each mode may be described by a state vector $|\ \rangle_k$ in a Hilbert space appropriate to that mode. The states of the entire field are then defined in the product space of the Hilbert spaces for all of the modes.

To discuss the quantum states of the individual modes we need only be familiar with the most elementary aspects of the treatment of a single harmonic oscillator. The Hamiltonian $\tfrac{1}{2}\hbar\omega_k(a_k^\dagger a_k + a_k a_k^\dagger)$ has eigenvalues $\hbar\omega_k(n_k + \tfrac{1}{2})$, where n_k is an integer ($n_k = 0, 1, 2 \cdots$). The state vector for the ground state of the oscillator will be written as $|\ \rangle_k$. It is defined by the condition

$$a_k |0\rangle_k = 0. \qquad (2.14)$$

The state vectors for the excited states of the oscillator may be obtained by applying integral powers of the operator $a_k^\dagger$ to $|0\rangle_k$. These states are written in normalized form as

$$|n_k\rangle_k = \frac{(a_k^\dagger)^{n_k}}{(n_k!)^{1/2}} |0\rangle_k, \quad (n_k = 0, 1, 2 \cdots). \qquad (2.15)$$

The way in which the operators a_k and $a_k^\dagger$ act upon these states is indicated by the relations

$$a_k |n_k\rangle_k = n_k^{1/2} |n_k - 1\rangle_k, \qquad (2.16)$$

$$a_k^\dagger |n_k\rangle = (n_k + 1)^{1/2} |n_k + 1\rangle_k, \qquad (2.17)$$

$$a_k^\dagger a_k |n_k\rangle = n_k |n_k\rangle. \qquad (2.18)$$

With these preliminaries completed we are now ready to discuss the coherent states of the field in greater detail. The expansion (2.10) for the vector potential exhibits its positive frequency part as the sum containing the photon annihilation operators a_k and its negative frequency part as that involving the creation operators $a_k^\dagger$. The positive frequency part of the electric field operator is thus given, according to (2.10), by

$$\mathbf{E}^{(+)}(\mathbf{r}t) = i \sum_k (\tfrac{1}{2}\hbar\omega_k)^{1/2} a_k \mathbf{u}_k(\mathbf{r}) e^{-i\omega_k t}. \qquad (2.19)$$

The eigenvalue functions $\boldsymbol{\mathcal{E}}(\mathbf{r}t)$ defined by Eq. (2.2) must clearly satisfy the Maxwell equations, just as the operator $\mathbf{E}^{(+)}(\mathbf{r}t)$ does. They therefore possess an expansion in normal modes similar to Eq. (2.19). In other words we may introduce a set of c-number Fourier coefficients α_k which permit us to write the eigenvalue function as

$$\boldsymbol{\mathcal{E}}(\mathbf{r}t) = i \sum_k (\tfrac{1}{2}\hbar\omega_k)^{1/2} \alpha_k \mathbf{u}_k(\mathbf{r}) e^{-i\omega_k t}. \qquad (2.20)$$

Since the mode functions $\mathbf{u}_k(\mathbf{r})$ form an orthogonal set, it then follows that the eigenstate $|\ \rangle$ for the field obeys the infinite succession of relations

$$a_k |\ \rangle = \alpha_k |\ \rangle, \qquad (2.21)$$

for all modes k. To find the states which satisfy these relations we seek states, $|\alpha_k\rangle_k$, of the individual modes which individually obey the relations

$$a_k |\alpha_k\rangle_k = \alpha_k |\alpha_k\rangle_k. \qquad (2.22)$$

The coherent states $|\ \rangle$ of the field, considered as a whole, are then seen to be direct products of the individual states $|\alpha_k\rangle$,

$$|\ \rangle = \prod_k |\alpha_k\rangle_k. \qquad (2.23)$$

III. COHERENT STATES OF A SINGLE MODE

The next few sections will be devoted to discussing the description of a single mode oscillator. We may therefore simplify the notation a bit by dropping the mode index k as a subscript to the state vector and to the amplitude parameters and operators. To find the oscillator state $|\alpha\rangle$ which satisfies

$$a|\alpha\rangle = \alpha|\alpha\rangle, \qquad (3.1)$$

we begin by taking the scalar product of both sides of the equation with the nth excited state, $\langle n|$. By using the Hermitian adjoint form of the relation (2.17), we find the recursion relation

$$(n+1)^{1/2}\langle n+1|\alpha\rangle = \alpha\langle n|\alpha\rangle \qquad (3.2)$$

for the scalar products $\langle n|\alpha\rangle$. We immediately find from the recursion relation that

$$\langle n|\alpha\rangle = \frac{\alpha^n}{(n!)^{1/2}}\langle 0|\alpha\rangle. \qquad (3.3)$$

These scalar products are the expansion coefficients of the state $|\alpha\rangle$ in terms of the complete orthonormal set $|n\rangle$ ($n = 0, 1, \cdots$). We thus have

$$|\alpha\rangle = \sum_n |n\rangle\langle n|\alpha\rangle$$

$$= \langle 0|\alpha\rangle \sum_n \frac{\alpha^n}{(n!)^{1/2}} |n\rangle. \qquad (3.4)$$

The squared length of the vector $|\alpha\rangle$ is thus

$$\langle \alpha|\alpha\rangle = |\langle 0|\alpha\rangle|^2 \sum_n \frac{|\alpha|^{2n}}{n!}$$

$$= |\langle 0|\alpha\rangle|^2 e^{|\alpha|^2}. \qquad (3.5)$$

If the state $|\alpha\rangle$ is normalized so that $\langle \alpha|\alpha\rangle = 1$ we may evidently define its phase by choosing

$$\langle 0|\alpha\rangle = e^{-\tfrac{1}{2}|\alpha|^2}. \qquad (3.6)$$

The coherent states of the oscillator therefore take the forms

$$|\alpha\rangle = e^{-\tfrac{1}{2}|\alpha|^2} \sum_n \frac{\alpha^n}{(n!)^{1/2}} |n\rangle \qquad (3.7)$$

and

$$\langle \alpha| = e^{-\tfrac{1}{2}|\alpha|^2} \sum_n \frac{(\alpha^*)^n}{(n!)^{1/2}} \langle n|. \qquad (3.8)$$

These forms show that the average occupation number of the nth state is given by a Poisson distribution with mean value $|\alpha|^2$,

$$|\langle n|\alpha\rangle|^2 = \frac{|\alpha|^{2n}}{n!} e^{-|\alpha|^2}. \qquad (3.9)$$

They also show that the coherent state $|\alpha\rangle$ corresponding to $\alpha = 0$ is the unique ground state of the oscillator, i.e., the state $|n\rangle$ for $n = 0$.

An alternative approach to the coherent states will also prove quite useful in the work to follow. For this purpose we assume that there exists a unitary operator D which acts as a displacement operator upon the amplitudes $a^\dagger$ and a. We let D be a function of a complex parameter β, and require that it displace the amplitude operators according to the scheme

$$D^{-1}(\beta) a D(\beta) = a + \beta, \qquad (3.10)$$

$$D^{-1}(\beta) a^\dagger D(\beta) = a^\dagger + \beta^*. \qquad (3.11)$$

Then if $|\alpha\rangle$ obeys Eq. (3.1), it follows that $D^{-1}(\beta)|\alpha\rangle$ is an eigenstate of a corresponding to the eigenvalue $\alpha-\beta$,

$$aD^{-1}(\beta)|\alpha\rangle=(\alpha-\beta)D^{-1}(\beta)|\alpha\rangle. \quad (3.12)$$

In particular, if we choose $\beta=\alpha$, we find

$$aD^{-1}(\alpha)|\alpha\rangle=0.$$

Since the ground state of the oscillator is uniquely defined by the relation (2.14), it follows that $D^{-1}(\alpha)|\alpha\rangle$ is just the ground state, $|0\rangle$. The coherent states, in other words, are just displaced forms of the ground state of the oscillator,

$$|\alpha\rangle=D(\alpha)|0\rangle. \quad (3.13)$$

To find an explicit form for the displacement operator $D(\alpha)$, we begin by considering infinitesimal displacements in the neighborhood of $D(0)=1$. For arbitrary displacements $d\alpha$, we see easily from the commutation rules (2.13) that $D(d\alpha)$ may be chosen to have the form

$$D(d\alpha)=1+a^\dagger d\alpha-ad\alpha^*, \quad (3.14)$$

which holds to first order in $d\alpha$. To formulate a simple differential equation obeyed by the unknown operator we consider increments of α of the form $d\alpha=\alpha d\lambda$ where λ is a real parameter. Then if we assume the operators D to possess the group multiplication property

$$D(\alpha(\lambda+d\lambda))=D(\alpha d\lambda)D(\alpha), \quad (3.15)$$

we find the differential equation

$$\frac{d}{d\lambda}D(\alpha\lambda)=(\alpha a^\dagger-\alpha^* a)D(\alpha\lambda), \quad (3.16)$$

whose solution, evaluated for $\lambda=1$, is the unitary operator

$$D(\alpha)=e^{\alpha a^\dagger-\alpha^* a}. \quad (3.17)$$

The coherent states $|\alpha\rangle$ may therefore be written in the form

$$|\alpha\rangle=e^{\alpha a^\dagger-\alpha^* a}|0\rangle \quad (3.18)$$

which is correctly normalized since $D(\alpha)$ is unitary.

It is interesting to discuss the relationship between the two forms we have derived for the coherent states. For this purpose we invoke a simple theorem on the multiplication of exponential functions of operators. If $\mathfrak{A}$ and $\mathfrak{B}$ are any two operators, whose commutator $[\mathfrak{A},\mathfrak{B}]$ commutes with each of them,

$$[[\mathfrak{A},\mathfrak{B}],\mathfrak{A}]=[[\mathfrak{A},\mathfrak{B}],\mathfrak{B}]=0, \quad (3.19)$$

it may be shown[4] that

$$\exp(\mathfrak{A})\exp(\mathfrak{B})=\exp\{\mathfrak{A}+\mathfrak{B}+\tfrac{1}{2}[\mathfrak{A},\mathfrak{B}]\}. \quad (3.20)$$

If we write $\mathfrak{A}=a^\dagger$ and $\mathfrak{B}=a$, this theorem permits us to resolve the exponential $D(\alpha)$ given by Eq. (3.17) into the product

$$D(\alpha)=e^{-\frac{1}{2}|\alpha|^2}e^{\alpha a^\dagger}e^{-\alpha^* a}. \quad (3.21)$$

Products of this type, which have been ordered so that the annihilation operators all stand to the right of the creation operators, will be said to be in normal form. Their convenience is indicated by the fact that the exponential $\exp[-\alpha^* a]$, when applied to the ground state $|0\rangle$, reduces in effect to unity, i.e., we have

$$e^{-\alpha^* a}|0\rangle=|0\rangle, \quad (3.22)$$

since the exponential may be expanded in series and the definition (2.14) of the ground state applied. It follows then that the coherent states may be written as

$$|\alpha\rangle=D(\alpha)|0\rangle$$
$$=e^{-\frac{1}{2}|\alpha|^2}e^{\alpha a^\dagger}|0\rangle \quad (3.23)$$
$$=e^{-\frac{1}{2}|\alpha|^2}\sum_n\frac{(\alpha a^\dagger)^n}{n!}|0\rangle. \quad (3.24)$$

Since the excited states of the oscillator are given by $|n\rangle=(n!)^{-1/2}(a^\dagger)^n|0\rangle$, we have once again derived the expression

$$|\alpha\rangle=e^{-\frac{1}{2}|\alpha|^2}\sum_n\frac{\alpha^n}{n!}|n\rangle.$$

It may help in visualizing the coherent states if we discuss the form they take in coordinate space and in momentum space. We therefore introduce a pair of Hermitian operators q and p to represent, respectively, the coordinate of the mode oscillator and its momentum. These operators, which must satisfy the canonical commutation relation, $[q,p]=i\hbar$, may be defined for our purposes by the familiar expressions

$$q=(\hbar/2\omega)^{1/2}(a^\dagger+a), \quad (3.25a)$$
$$p=i(\hbar\omega/2)^{1/2}(a^\dagger-a). \quad (3.25b)$$

To find the expectation value of q and p in the coherent states we need only use Eq. (3.1), which defines these states, and its corresponding Hermitian adjoint form. We have then

$$\langle\alpha|q|\alpha\rangle=(2\hbar/\omega)^{1/2}\operatorname{Re}\alpha, \quad (3.26a)$$
$$\langle\alpha|p|\alpha\rangle=(2\hbar\omega)^{1/2}\operatorname{Im}\alpha, \quad (3.26b)$$

where $\operatorname{Re}\alpha$ and $\operatorname{Im}\alpha$ stand for the real and imaginary parts of α.

To find the wave functions for the coherent states, we write the defining equation (3.1) in the form

$$(2\hbar\omega)^{-1/2}(\omega q+ip)|\alpha\rangle=\alpha|\alpha\rangle, \quad (3.27)$$

and take the scalar product of both members with the conjugate state $\langle q'|$, which corresponds to the eigenvalue q' for q. Since the momentum may be represented by a derivative operator, i.e., $\langle q'|p=-i\hbar(d/dq')\langle q'|$, we find that the coordinate space wave function, $\langle q'|\alpha\rangle$,

---

[4] A. Messiah, *Quantum Mechanics* (North-Holland Publishing Company, Amsterdam, 1961), Vol. I, p. 442.

obeys the differential equation

$$\frac{d}{dq'}\langle q'|\alpha\rangle = -2\left(\frac{\omega}{2\hbar}\right)^{1/2}\left\{\left(\frac{\omega}{2\hbar}\right)^{1/2}q' - \alpha\right\}\langle q'|\alpha\rangle. \quad (3.28)$$

The equation may be integrated immediately to yield a solution for the wave function which, in normalized form, is

$$\langle q'|\alpha\rangle = (\omega/\pi\hbar)^{1/4}\exp\{-[(\omega/2\hbar)^{1/2}q' - \alpha]^2\}. \quad (3.29)$$

An analogous argument furnishes the momentum space wave function. If we take the scalar product of Eq. (3.27) with a momentum eigenstate $\langle p'|$, and use the relation $\langle p'|q = i\hbar(\partial/\partial p')\langle p'|$, we reach a differential equation whose normalized solution is

$$\langle p'|\alpha\rangle = (\pi\hbar\omega)^{-1/4}\exp\{-[(2\hbar\omega)^{-1/2}p' + i\alpha]^2\}. \quad (3.30)$$

Both of these wave functions are simply displaced forms of the ground-state wave function for the oscillator. The parameters $(\hbar/\omega)^{1/2}$ and $(\hbar\omega)^{1/2}$ correspond to the amplitudes of the zero-point fluctuations of the coordinate and momentum, respectively, for an oscillator of unit mass. The fact that the wave functions for the coherent states have this elementary structure should be no surprise in view of the way they are generated in Eq. (3.13), by means of displacements in the complex α plane.

The time-independent states $|\alpha\rangle$ which we have been describing are those characteristic of the Heisenberg picture of quantum mechanics. The Schrödinger picture, alternatively, would make use of the time-dependent states $\exp(-iHt/\hbar)|\alpha\rangle$. If we omit the zero-point energy $\frac{1}{2}\hbar\omega$ from the oscillator Hamiltonian and write $H = \hbar\omega a^\dagger a$, it is then clear from the expansion (3.7) for $|\alpha\rangle$ that the corresponding Schrödinger state takes the same form with α replaced by $\alpha e^{-i\omega t}$. We may thus write the Schrödinger state as $|\alpha e^{-i\omega t}\rangle$. With the substitution of $\alpha e^{-i\omega t}$ for α in Eqs. (3.26a) and (3.26b), we see that the expectation values of the coordinate and momentum carry out a simple harmonic motion with coordinate amplitude $(2\hbar/\omega)^{1/2}|\alpha|$. The same substitutions in the wave functions (3.29) and (3.30) show that the Gaussian probability densities characteristic of the ground state of the oscillator are simply carried back and forth in the same motion as the expectation values. Such wave packets are, of course, quite familiar; they were introduced to quantum mechanics at a very early stage by Schrödinger,[5] and have often been used to illustrate the way in which the behavior of the oscillator approaches the classical limit.

Another connection in which the wave packets (3.29) and (3.30) have been discussed in the past has to do with the particular way in which they localize the coordinate q' and the momentum p'. Wave packets can, of course, be found which localize either variable more sharply, but only at the expense of the localization of the other. There is a sense in which the wave packets (3.29) and (3.30) furnish a unique compromise; they minimize the product of the uncertainties of the variables q' and p'. If we represent expectation values by means of the angular brackets $\langle\ \rangle$ and define the variances

$$(\Delta q)^2 = \langle q^2\rangle - \langle q\rangle^2, \quad (3.31a)$$
$$(\Delta p)^2 = \langle p^2\rangle - \langle p\rangle^2, \quad (3.31b)$$

we find, for the wave functions (3.29) and (3.30), that the product of the variances is

$$(\Delta p)^2(\Delta q)^2 = \tfrac{1}{4}\hbar^2.$$

According to the uncertainty principle, this is the minimum value such a product can have.[6] There thus exists a particular sense in which the description of an oscillator by means of the wave functions (3.29) and (3.30) represents as close an approach to classical localization as is possible.

The uses we shall make of the coherent states in quantum electrodynamics will not, in fact, require the explicit introduction of coordinate or momentum variables. We have reviewed the familiar representations of the coherent states in terms of these variables in the hope that they may be of some help in understanding the various applications of the states which we shall shortly undertake.

One property of the states $|\alpha\rangle$ which is made clear by the wave-function representations is that two such states are not, in general, orthogonal to one another. If we consider, for example, the wave functions $\langle q'|\alpha\rangle$ and $\langle q'|\alpha'\rangle$ for values of α' close to α, it is evident that the functions are similar in form and overlap one another appreciably. For values of α' quite different from α, however, the overlap is at most quite small. We may therefore expect that the scalar product $\langle\alpha|\alpha'\rangle$, which is unity for $\alpha'=\alpha$, will tend to decrease in absolute magnitude as α' and α recede from one another in the complex plane. The scalar product may, in fact, be calculated more simply than by using wave functions if we employ the representations (3.7) and (3.8). We then find

$$\langle\alpha|\beta\rangle = e^{-\frac{1}{2}|\alpha|^2 - \frac{1}{2}|\beta|^2}\sum_{n,m}\frac{(\alpha^*)^n\beta^m}{(n!m!)^{1/2}}\langle n|m\rangle,$$

which, in view of the orthonormality of the $|n\rangle$ states, reduces to

$$\langle\alpha|\beta\rangle = \exp\{\alpha^*\beta - \tfrac{1}{2}|\alpha|^2 - \tfrac{1}{2}|\beta|^2\}. \quad (3.32)$$

The absolute magnitude of the scalar product is given by

$$|\langle\alpha|\beta\rangle|^2 = \exp\{-|\alpha-\beta|^2\}, \quad (3.33)$$

---

[5] E. Schrödinger, Naturwissenschaften **14**, 664 (1926). For a more recent treatment see L. I. Schiff, *Quantum Mechanics* (McGraw-Hill Book Company, Inc., New York, 1955), 2nd ed., p. 67.

[6] W. Heisenberg, *The Physical Principles of the Quantum Theory* (University of Chicago Press, Chicago, 1930, reprinted by Dover Publications, Inc., New York, 1930), pp. 16–19.

which shows that the coherent states tend to become approximately orthogonal for values of α and β which are sufficiently different. The fact that these states are not even approximately orthogonal for $|\alpha-\beta|$ of order unity may be regarded as an expression of the overlap caused by the presence of the displaced zero-point fluctuations.

Since the coherent states do not form an orthogonal set, they appear to have received little attention as a possible system of basis vectors for the expansion of arbitrary states.[7] We shall show in the following section that such expansions can be carried out conveniently and uniquely and that they possess exceedingly useful properties. In later sections we shall, by generalizing the procedure to deal with bilinear combinations of states $|\alpha\rangle$ and $\langle\beta|$, develop analogous expansions for operators[1] as well.

IV. EXPANSION OF ARBITRARY STATES IN TERMS OF COHERENT STATES

While orthogonality is a convenient property for a set of basis states it is not a necessary one. The essential property of such a set is that it be complete. The set of coherent states $|\alpha\rangle$ for a mode oscillator can be shown without difficulty to form a complete set. To give a proof we need only demonstrate that the unit operator may be expressed as a suitable sum or an integral, over the complex α plane, of projection operators of the form $|\alpha\rangle\langle\alpha|$. In order to describe such integrals we introduce the differential element of area in the α plane

$$d^2\alpha = d(\text{Re } \alpha) d(\text{Im } \alpha) \quad (4.1)$$

(i.e., $d^2\alpha$ is real). If we write $\alpha = |\alpha| e^{i\vartheta}$, we may easily prove the integral identity

$$\int (\alpha^*)^n \alpha^m e^{-|\alpha|^2} d^2\alpha$$
$$= \int_0^\infty |\alpha|^{n+m+1} e^{-|\alpha|^2} d|\alpha| \int_0^{2\pi} e^{i(m-n)\vartheta} d\vartheta$$
$$= \pi n! \delta_{nm}, \quad (4.2)$$

in which the integration is carried out, as indicated, over the entire area of the complex plane. With the aid of this identity and the expansions (3.7,8) for the coherent states, we may immediately show

$$\int |\alpha\rangle\langle\alpha| d^2\alpha = \pi \sum_n |n\rangle\langle n|.$$

Since the n-quantum states are known to form a complete orthonormal set, the indicated sum over n is simply the unit operator. We have thus shown[1]

$$\frac{1}{\pi} \int |\alpha\rangle\langle\alpha| d^2\alpha = 1, \quad (4.3)$$

which is a completeness relation for the coherent states of precisely the type desired.

An arbitrary state of an oscillator must possess an expansion in terms of the n-quantum states of the form

$$|\ \rangle = \sum_n c_n |n\rangle,$$
$$= \sum_n c_n \frac{(a^\dagger)^n}{(n!)^{1/2}} |0\rangle, \quad (4.4)$$

where $\sum |c_n|^2 = 1$. The series which occurs in Eq. (4.4) may be used to define a function f of a complex variable z,

$$f(z) = \sum c_n \frac{z^n}{(n!)^{1/2}}. \quad (4.5)$$

It is clear from the normalization condition on the c_n that this series converges for all finite z, and thus represents a function which is analytic throughout the finite complex plane. We shall speak of the functions $f(z)$ for which $\sum |c_n|^2 = 1$ as the set of normalized entire functions. There is evidently a one-to-one correspondence which exists between such entire functions and the states of the oscillator. One way of approaching the description of the oscillator is to regard the functions $f(z)$ themselves as the elements of a Hilbert space. The properties of this space and of expansions carried out in it have been studied in some detail by Segal[8] and Bargmann.[9] The method we shall use for expanding arbitrary states in terms of the coherent states has been developed as a simple generalization of the usual method for carrying out changes of basis states in quantum mechanics. It is evidently equivalent, however, to one of the expansions stated by Bargmann.

If we designate the arbitrary state which corresponds to the function $f(z)$ by $|f\rangle$, then we may rewrite Eq. (4.4) as

$$|f\rangle = f(a^\dagger) |0\rangle. \quad (4.6)$$

To secure the expansion of $|f\rangle$ in terms of the states $|\alpha\rangle$, we multiply $|f\rangle$ by the representation (4.3) of the unit operator. We then find

$$|f\rangle = \frac{1}{\pi} \int |\alpha\rangle\langle\alpha| f(a^\dagger) |0\rangle d^2\alpha,$$

---

[7] Uses of these states as generating functions for the n-quantum states have, however, been made by J. Schwinger, Phys. Rev. **91**, 728 (1953).

[8] I. E. Segal, Illinois J. Math. **6**, 520 (1962).
[9] V. Bargmann, Commun. Pure and Appl. Math. **14**, 187 (1961); Proc. Natl. Acad. Sci. U. S. **48**, 199 (1962).

which reduces, since $\langle \alpha | f(a^\dagger) = \langle \alpha | f(\alpha^*)$, to

$$|f\rangle = \frac{1}{\pi} \int |\alpha\rangle f(\alpha^*) e^{-\frac{1}{2}|\alpha|^2} d^2\alpha, \qquad (4.7)$$

which is an expansion of the desired type.

It is worth noting that the expansion (4.7) can easily be inverted to furnish an explicit form for the function $f(\alpha^*)$ which corresponds to any vector $|f\rangle$. For this purpose we take the scalar product of both sides of Eq. (4.7) with the coherent state $\langle \beta |$, and then, using Eq. (3.32), evaluate the scalar product $\langle \beta | \alpha \rangle$ to find

$$\langle \beta | f \rangle = \frac{1}{\pi} e^{-\frac{1}{2}|\beta|^2} \int e^{\beta^*\alpha - |\alpha|^2} f(\alpha^*) d^2\alpha. \qquad (4.8)$$

Since $f(\alpha^*)$ may be expanded in a convergent power series we note the relation

$$\frac{1}{\pi} \int e^{\beta^*\alpha - |\alpha|^2} (\alpha^*)^n d^2\alpha = (\beta^*)^n, \qquad (4.9)$$

from which we may derive the more general identity

$$\frac{1}{\pi} \int e^{\beta^*\alpha - |\alpha|^2} f(\alpha^*) d^2\alpha = f(\beta^*). \qquad (4.10)$$

On substituting the latter identity in Eq. (4.8) we find

$$f(\beta^*) = e^{\frac{1}{2}|\beta|^2} \langle \beta | f \rangle. \qquad (4.11)$$

There is thus a unique correspondence between functions $f(\alpha^*)$ which play the role of expansion amplitudes in Eq. (4.7) and the vectors $|f\rangle$ which describe the state of the oscillator.

An expansion analogous to Eq. (4.7) also exists for the adjoint state vectors. If we let $g(\alpha^*)$ be an entire function of α^* we may construct for the state $\langle g |$ the expansion

$$\langle g | = \frac{1}{\pi} \int [g(\beta^*)]^* \langle \beta | e^{-\frac{1}{2}|\beta|^2} d^2\beta. \qquad (4.12)$$

The scalar product of the two states $\langle g |$ and $|f\rangle$ may then be expressed as

$$\langle g | f \rangle = \pi^{-2} \int [g(\beta^*)]^* f(\alpha^*) \exp\{\beta^*\alpha - |\alpha|^2 - |\beta|^2\} d^2\alpha d^2\beta.$$

The identity (4.10) permits us to carry out the integration over the variable α to find

$$\langle g | f \rangle = \frac{1}{\pi} \int [g(\beta^*)]^* f(\beta^*) e^{-|\beta|^2} d^2\beta. \qquad (4.13)$$

This expression for the scalar product of two vectors is, in essence, the starting point used by Bargmann in his discussion[10] of the Hilbert space of functions $f(z)$.

[10] Some of Bargmann's arguments are summarized by S. Schweber, J. Math. Phys. **3**, 831 (1962), who has used them in

It may be worth noting, for its mathematical interest, that the coherent states $|\alpha\rangle$ are not linearly independent of one another, as the members of a complete orthogonal set would be. Thus, for example, the expansion (4.7) may be used to express any given coherent state linearly in terms of all of the others, i.e., in view of Eqs. (4.11) and (3.32) we may write

$$|\alpha\rangle = \frac{1}{\pi} \int |\beta\rangle e^{\beta^*\alpha - \frac{1}{2}|\alpha|^2 - \frac{1}{2}|\beta|^2} d^2\beta. \qquad (4.14)$$

There exist many other types of linear dependence among the states $|\alpha\rangle$. We may, for example, note the identity

$$\int |\alpha\rangle \alpha^n e^{-\frac{1}{2}|\alpha|^2} d^2\alpha = 0, \qquad (4.15)$$

which holds for all integral $n>0$. It is clear from the latter result that if we admitted as expansion coefficients in Eq. (4.7) more general functions than $f(\alpha^*)$, say functions $F(\alpha,\alpha^*)$, there would be many additional ways of expanding any state in terms of coherent states. The constraint implicit in Eq. (4.7), that the expansion function must depend analytically upon the variable α^* is what renders the expansion unique. The virtue of an expansion scheme in which the coefficients are uniquely determined is evident. It becomes possible, by inverting the expansion as in Eq. (4.11), to construct an explicit solution for the expansion coefficient of any state, no matter what representation it was expressed in initially.

V. EXPANSION OF OPERATORS IN TERMS OF COHERENT STATE VECTORS

Our knowledge of the condition of an oscillator mode is rarely explicit enough in practice to permit the specification of its quantum state. Instead, we must describe it in terms of a mixture of states which is expressed by means of a density operator. The same reasons that lead us to express arbitrary states in terms of the coherent states, therefore, suggest that we develop an expansion for the density operator in terms of these states as well. We shall begin by considering in the present section a rather more general class of operators and then specialize to the case of the density operator in the section which follows.

A general quantum mechanical operator T may be expressed in terms of its matrix elements connecting states with fixed numbers of quanta as

$$T = \sum_{n,m} |n\rangle T_{nm} \langle m |, \qquad (5.1)$$

$$= \sum T_{nm} (n!m!)^{-1/2} (a^\dagger)^n |0\rangle \langle 0| a^m. \qquad (5.2)$$

connection with the formulation of quantum mechanics in terms of Feynman amplitudes. We are indebted to Dr. S. Bergmann for calling this reference to our attention.

If we use this expression for T to calculate the matrix element which connects the two coherent states $\langle\alpha|$ and $\langle\beta|$ we find

$$\langle\alpha|T|\beta\rangle = \sum_{n,m} T_{nm}(n!m!)^{-1/2}(\alpha^*)^n\beta^m\langle\alpha|0\rangle\langle0|\beta\rangle. \quad (5.3)$$

It is evidently convenient to define a function $\mathcal{T}(\alpha^*,\beta)$ as

$$\mathcal{T}(\alpha^*,\beta) = \sum_{n,m} T_{nm}(n!m!)^{-1/2}(\alpha^*)^n\beta^m. \quad (5.4)$$

The operators which occur in quantum mechanics are often unbounded ones such as those of Eqs. (2.16)–(2.18). Those operators and the others we are apt to encounter have the property that the magnitudes of the matrix elements T_{nm} are dominated by an expression of the form Mn^jm^k for some fixed positive values of M, j, and k. It then follows that the double series (5.4) converges throughout the finite α^* and β planes and represents an entire function of both variables.

To secure the expansion of the operator T in terms of the coherent states, we may use the representation (4.3) of the unit operator to write

$$T = \frac{1}{\pi^2}\int |\alpha\rangle\langle\alpha|T|\beta\rangle\langle\beta|d^2\alpha d^2\beta, \quad (5.5)$$

$$= \frac{1}{\pi^2}\int |\alpha\rangle \mathcal{T}(\alpha^*,\beta)\langle\beta|\langle\alpha|0\rangle\langle0|\beta\rangle d^2\alpha d^2\beta,$$

$$= \frac{1}{\pi^2}\int |\alpha\rangle \mathcal{T}(\alpha^*,\beta)\langle\beta|\exp\{-\tfrac{1}{2}|\alpha|^2-\tfrac{1}{2}|\beta|^2\}d^2\alpha d^2\beta. \quad (5.6)$$

The inversion of this expansion, or the solution for $\mathcal{T}(\alpha^*,\beta)$, is accomplished by the same method we used to invert Eq. (4.7) and secure the amplitude function (4.11). The result of the inversion is

$$\mathcal{T}(\alpha^*,\beta) = \langle\alpha|T|\beta\rangle\exp\{\tfrac{1}{2}|\alpha|^2+\tfrac{1}{2}|\beta|^2\}. \quad (5.7)$$

We see, thus, that the expansion of operators, as well as of arbitrary quantum states, in terms of the coherent states is a unique one.

The law of operator multiplication is easily expressed in terms of the functions $\mathcal{T}$. If $T = T_1T_2$ and $\mathcal{T}_1$ and $\mathcal{T}_2$ are the functions appropriate to the latter two operators, we note that

$$\langle\alpha|T|\beta\rangle = \langle\alpha|T_1T_2|\beta\rangle$$

$$= \frac{1}{\pi}\int\langle\alpha|T_1|\gamma\rangle\langle\gamma|T_2|\beta\rangle d^2\gamma. \quad (5.8)$$

The function $\mathcal{T}$ which represents the product is therefore given by

$$\mathcal{T}(\alpha^*,\beta) = \frac{1}{\pi}\int \mathcal{T}_1(\alpha^*,\gamma)\mathcal{T}_2(\gamma^*,\beta)e^{-|\gamma|^2}d^2\gamma. \quad (5.9)$$

The expansion function for the operator $T^\dagger$, the Hermitian adjoint of T, is obtained by substituting T_{mn}^* for T_{nm} in Eq. (5.4). It is given by $[\mathcal{T}(\beta^*,\alpha)]^*$. If the operator T is Hermitian the function $\mathcal{T}$ must satisfy the identity

$$\mathcal{T}(\alpha^*,\beta) = [\mathcal{T}(\beta^*,\alpha)]^*, \quad (5.10)$$

since the expansions of T and $T^\dagger$ are unique.

The functions $\mathcal{T}(\alpha^*,\beta)$ which represent normal products of the operators $a^\dagger$ and a such as $(a^\dagger)^n a^m$ are immediately seen from Eqs. (5.7) and (3.32) to be

$$\mathcal{T}(\alpha^*,\beta) = (\alpha^*)^n\beta^m \exp[\alpha^*\beta]. \quad (5.11)$$

In particular, the unit operator corresponds to $n=m=0$. It may be worth noting at this point that many of the foregoing formulas can be abbreviated somewhat by adopting a normalization different from the conventional one for the coherent states. If we introduce the symbol $\|\alpha\rangle$ for the states normalized in the new way and define these as

$$\|\alpha\rangle = |\alpha\rangle e^{\frac{1}{2}|\alpha|^2}, \quad (5.12)$$

then we may write the scalar product of two such states as $\langle\alpha\|\beta\rangle$. We see from Eq. (3.32) that this scalar product is

$$\langle\alpha\|\beta\rangle = \exp[\alpha^*\beta]. \quad (5.13)$$

We may next, following Bargmann,[9] introduce an element of measure $d\mu(\alpha)$ which is defined as

$$d\mu(\alpha) = \frac{1}{\pi}e^{-|\alpha|^2}d^2\alpha. \quad (5.14)$$

With these alterations, all of the Gaussian functions, and factors of π, in the preceding formulas become absorbed, as it were, into the notation. The Eqs. (5.6) and (5.7), for example, reduce to the briefer forms

$$T = \int \|\alpha\rangle \mathcal{T}(\alpha^*,\beta)\langle\beta\| d\mu(\alpha)d\mu(\beta) \quad (5.15)$$

and

$$\mathcal{T}(\alpha^*,\beta) = \langle\alpha\|T\|\beta\rangle. \quad (5.16)$$

A more significant property of the states $\|\alpha\rangle$ is that they are given by the expansion

$$\|\alpha\rangle = \sum_n \frac{\alpha^n}{(n!)^{1/2}}|n\rangle \quad (5.17)$$

and thus obey the relation

$$a^\dagger\|\alpha\rangle = \frac{\partial}{\partial\alpha}\|\alpha\rangle. \quad (5.18)$$

While the properties of the alternatively normalized states $\|\alpha\rangle$ are worth bearing in mind, we have chosen not to adopt this normalization in the present paper in order to retain the more conventional interpretation of scalar products as probability amplitudes. The advantage afforded by the relation (5.18) is not a great one since all of the operators we shall have to deal with are either already in normally ordered form, or easily so ordered.

VI. GENERAL PROPERTIES OF THE DENSITY OPERATOR

The formalism we have developed in the two preceding sections has been intended to provide a background for the expression of the density operator of a mode in terms of the vectors that represent coherent states. Viewed in mathematical terms, the use of the coherent state vectors in this way leads to considerable simplification in the calculation of statistical averages. The fact that these states are eigenstates of the field operators $\mathbf{E}^{(\pm)}(rt)$ means that normally ordered products of the field operators, when they are to be averaged, may be replaced by the products of their eigenvalues, i.e., treated not as operators, but as numbers. The field correlation functions such as $G^{(1)}$ given by Eq. (2.1) are averages of just such operator products. Their evaluation may be carried out quite conveniently through use of the representations we shall discuss.

Any density operator ρ may, according to the methods of the preceding section, be represented in a unique way by means of a function of two complex variables, $R(\alpha^*,\beta)$, which is analytic throughout the finite α^* and β planes. The function R is given explicitly, by means of Eq. (5.7), as

$$R(\alpha^*,\beta) = \langle \alpha | \rho | \beta \rangle \exp[\tfrac{1}{2}|\alpha|^2 + \tfrac{1}{2}|\beta|^2]. \quad (6.1)$$

If we happen to know the matrix representation of ρ in the basis formed by the n-quantum states, the function R is evidently given by

$$R(\alpha^*,\beta) = \sum_{n,m} \langle n | \rho | m \rangle (n!m!)^{-1/2} (\alpha^*)^n \beta^m. \quad (6.2)$$

If we do not know the matrix elements $\langle n | \rho | m \rangle$ they may be found quite simply from a knowledge of $R(\alpha^*,\beta)$. One method for finding them is to consider $R(\alpha^*,\beta)$ as a generating function and identify its Taylor series with the series (6.2). A second method is to note that if we multiply Eq. (6.2) by $\alpha^i (\beta^*)^j \exp[-(|\alpha|^2 + |\beta|^2)]$ and integrate over the α and β planes, then all terms save that for $n=i$ and $m=j$ vanish in the sum on the right and we have

$$\langle i | \rho | j \rangle = \frac{1}{\pi^2} \int R(\alpha^*,\beta)(i!j!)^{-1/2} \alpha^i (\beta^*)^j e^{-(|\alpha|^2+|\beta|^2)} d^2\alpha d^2\beta. \quad (6.3)$$

Given the knowledge of $R(\alpha^*,\beta)$, we may write the density operator as

$$\rho = \frac{1}{\pi^2} \int |\alpha\rangle R(\alpha^*,\beta) \langle\beta| e^{-\tfrac{1}{2}(|\alpha|^2+|\beta|^2)} d^2\alpha d^2\beta. \quad (6.4)$$

The statistical average of an operator T is given by the trace of the product ρT. If we calculate this average by using the representation (6.4) for ρ we must note that the trace of the expression $|\alpha\rangle\langle\beta|T$, regarded as an operator, is the matrix element $\langle\beta|T|\alpha\rangle$. Then, if we express the matrix element in terms of the function $\mathcal{T}(\alpha^*,\beta)$ defined by Eq. (5.7) we find

$$\operatorname{tr}\{\rho T\} = \frac{1}{\pi^2} \int R(\alpha^*,\beta) \mathcal{T}(\beta^*,\alpha) e^{-|\alpha|^2-|\beta|^2} d^2\alpha d^2\beta. \quad (6.5)$$

If T is any operator of the form $(a^\dagger)^n a^m$, its representation $\mathcal{T}(\beta^*,\alpha)$ is given by Eq. (5.11). In particular for $n=m=0$, we have the unit operator $T=1$ which is represented by $\mathcal{T}(\beta^*,\alpha) = \exp[\beta^*\alpha]$. Hence, the trace of ρ itself, which must be normalized to unity, is

$$\operatorname{tr}\rho = 1$$
$$= \frac{1}{\pi^2} \int R(\alpha^*,\beta) \exp[\beta^*\alpha - |\alpha|^2 - |\beta|^2] d^2\alpha d^2\beta.$$

Since $R(\alpha^*,\beta)$ is an entire function of α^*, we may use Eq. (4.10) to carry out the integration over the α plane. In this way we see that the normalization condition on R is

$$\frac{1}{\pi} \int R(\beta^*,\beta) e^{-|\beta|^2} d^2\beta = 1. \quad (6.6)$$

The density operator is Hermitian and hence has real eigenvalues. These eigenvalues may be interpreted as probabilities and so must be positive numbers. Since ρ is thus a positive definite operator, its expectation value in any state, e.g., the state $|f\rangle$ defined by Eq. (4.6), must be non-negative,

$$\langle f | \rho | f \rangle \geq 0. \quad (6.7)$$

If, for example, we choose the state $|f\rangle$ to be a coherent state $|\alpha\rangle$ we find that the function R, which is given by Eq. (6.1), satisfies the inequality

$$R(\alpha^*,\alpha) \geq 0. \quad (6.8)$$

If we let the state $|f\rangle$ be specified as in Eq. (4.7) by an entire function $f(\alpha^*)$, then we find from the inequality (6.7) the more general condition for positive definiteness

$$\int [f(\alpha^*)]^* f(\beta^*) R(\alpha^*,\beta) e^{-|\alpha|^2-|\beta|^2} d^2\alpha d^2\beta \geq 0, \quad (6.9)$$

which must hold for all entire functions f.

In many types of physical experiments, particularly those dealing with fields which oscillate at extremely high frequencies, we cannot be said to have any *a priori* knowledge of the time-dependent parameters. The predictions we make in such circumstances are unchanged by displacements in time. They may be derived from a density operator which is stationary, that is, one which commutes with the Hamiltonian operator or, more simply, with $a^\dagger a$. The necessary and sufficient condition that a function $R(\alpha^*,\beta)$ correspond to a stationary density operator is that it depend only on the product of its two variables, $\alpha^*\beta$. There must, in other words, exist an analytic function $\mathcal{S}$ such that

$$R(\alpha^*,\beta) = \mathcal{S}(\alpha^*\beta). \tag{6.10}$$

That this condition is a sufficient one is clear from the invariance of R under the multiplication of both α and β by a phase factor, $e^{i\varphi}$. The condition may be derived as a necessary one directly from the vanishing of the commutator of ρ with $a^\dagger a$. An alternative and perhaps simpler way of seeing the result depends on noting that a stationary ρ can only be a function of the Hamiltonian for the mode, or of $a^\dagger a$. It is therefore diagonal in the basis formed by the n-quantum states, i.e., $\langle n|\rho|m\rangle = \delta_{nm}\langle n|\rho|n\rangle$. Examination of the series expansion (6.2) for R then shows that it then takes the form of Eq. (6.10).

VII. THE P REPRESENTATION OF THE DENSITY OPERATOR

In the preceding sections we have demonstrated the generality of the use of the coherent states as a basis. Not all fields require for their description density operators of quite so general a form. Indeed for a broad class of radiation fields which includes, as we shall see, virtually all of those studied in optics, it becomes possible to reduce the density operator to a considerably simpler form. This form is one which brings to light many similarities between quantum electrodynamical calculations and the corresponding classical ones. Its use offers deep insights into the reasons why some of the fundamental laws of optics, such as those for superposition of fields and calculation of the resulting intensities, are the same as in classical theory, even when very few quanta are involved. We shall continue, for the present, to limit consideration to a single mode of the field.

One type of oscillator state which interests us particularly is, of course, a coherent state. The density operator for a pure state $|\alpha\rangle$ is just the projection operator

$$\rho = |\alpha\rangle\langle\alpha|. \tag{7.1}$$

The unique representation of this operator as a function $R(\beta^*,\gamma)$ is easily shown, from Eq. (6.1), to be

$$R(\beta^*,\gamma) = \exp[\beta^*\alpha + \gamma\alpha^* - |\alpha|^2]. \tag{7.2}$$

Other functions $R(\beta^*,\gamma)$, which satisfy the analyticity requirements necessary for the representations of density operators, may be constructed by forming linear combinations of exponentials such as (7.2) for various values of the complex parameter α. The functions R, which we form in this way, represent statistical mixtures of the coherent states. The most general such function R may be written as

$$R(\beta^*,\gamma) = \int P(\alpha) \exp[\beta^*\alpha + \gamma\alpha^* - |\alpha|^2] d^2\alpha, \tag{7.3}$$

where $P(\alpha)$ is a weight function defined at all points of the complex α plane. Since $R(\beta^*,\gamma)$ must satisfy the Hermiticity condition, Eq. (5.10), we require that the weight function be real-valued, i.e., $[P(\alpha)]^* = P(\alpha)$. The function $P(\alpha)$ need not be subject to any regularity conditions, but its singularities must be integrable ones.[11] It is convenient to allow $P(\alpha)$ to have delta-function singularities so that we may think of a pure coherent state as represented by a special case of Eq. (7.3). A real-valued two-dimensional delta function which is suited to this purpose may be defined as

$$\delta^{(2)}(\alpha) = \delta(\text{Re}\,\alpha)\delta(\text{Im}\,\alpha). \tag{7.4}$$

The pure coherent state $|\beta\rangle$ is then evidently described by

$$P(\alpha) = \delta^{(2)}(\alpha - \beta), \tag{7.5}$$

and the ground state of the oscillator is specified by setting $\beta = 0$.

The density operator ρ which corresponds to Eq. (7.3) is just a superposition of the projection operators (7.1),

$$\rho = \int P(\alpha) |\alpha\rangle\langle\alpha| d^2\alpha. \tag{7.6}$$

It is the kind of operator we might naturally be led to if we were given knowledge that the oscillator is in a coherent state, but one which corresponds to an unknown eigenvalue α. The function $P(\alpha)$ might then be thought of as playing a role analogous to a probability density for the distribution of values of α over the complex plane.[12] Such an interpretation may, as we shall see, be justified at times. In general, however, it is not possible to interpret the function $P(\alpha)$ as a probability distribution in any precise way since the projection operators $|\alpha\rangle\langle\alpha|$ with which it is associated are not orthogonal to one another for different values of α. There is an approximate sense, as we have noted in connection with Eq. (3.33), in which two states $|\alpha\rangle$ and $|\alpha'\rangle$ may be said to become orthogonal to one another for $|\alpha - \alpha'| \gg 1$, i.e., when their wave packets (3.29) and those of the form (3.30) do not appreciably overlap. When the function $P(\alpha)$ tends to vary little over such large ranges of the parameter α, the non-orthogonality of the coherent states will make little difference, and $P(\alpha)$ will then be interpretable approximately as a probability density. The functions $P(\alpha)$

---
[11] If the singularities of $P(\alpha)$ are of types stronger than those of delta functions, e.g., derivatives of delta functions, the field represented will have no classical analog.
[12] The existence of this form for the density operator has also been observed by E. C. G. Sudarshan, Phys. Rev. Letters **10**, 277 (1963). His note is discussed briefly at the end of Sec. X.

which vary this slowly will, in general, be associated with strong fields, ones which may be described approximately in classical terms.

We shall call the expression (7.6) for the density operator the P representation in order to distinguish it from the more general form based on the functions R discussed earlier. The normalization property of the density operator requires that $P(\alpha)$ obey the normalization condition

$$\mathrm{tr}\rho = \int P(\alpha) d^2\alpha = 1. \qquad (7.7)$$

It is interesting to examine the conditions that the positive definiteness of ρ places upon $P(\alpha)$. If we apply the condition (6.9) to the function $R(\beta^*,\gamma)$ given by Eq. (7.3) we find

$$\int [f(\beta^*)]^* f(\gamma^*) P(\alpha) \exp[\beta^*\alpha + \gamma\alpha^* - |\alpha|^2 - |\beta|^2 - |\gamma|^2]$$

$$\times d^2\alpha d^2\beta d^2\gamma \geq 0. \qquad (7.8)$$

The γ integration may be carried out via Eq. (4.10) and the β integration by means of its complex conjugate. We then have the condition that

$$\int |f(\alpha^*)|^2 P(\alpha) e^{-|\alpha|^2} d^2\alpha \geq 0 \qquad (7.9)$$

must hold for all entire functions $f(\alpha^*)$. In particular, the choice $f(\alpha^*) = \exp[\beta\alpha^* - \tfrac{1}{2}|\beta|^2]$ leads to the simple condition

$$\int P(\alpha) e^{-|\alpha - \beta|^2} d^2\alpha \geq 0, \qquad (7.10)$$

which must hold for all complex values of β. It corresponds to the requirement $\langle \beta | \rho | \beta \rangle \geq 0$. These conditions are immediately satisfied if $P(\alpha)$ is positive valued as it would be, were it a probability density. They are not strong enough, however, to exclude the possibility that $P(\alpha)$ takes on negative values over some suitably restricted regions of the plane.[13] This result serves to underscore the fact that the weight function $P(\alpha)$ cannot, in general, be interpreted as a probability density.[14]

If a density operator is specified by means of the P representation, its matrix elements connecting the n-quantum states are given by

$$\langle n | \rho | m \rangle = \int P(\alpha) \langle n | \alpha \rangle \langle \alpha | m \rangle d^2\alpha. \qquad (7.11)$$

When Eqs. (3.3) and (3.6) are used to evaluate the scalar products in the integrand we find

$$\langle n | \rho | m \rangle = (n!m!)^{-1/2} \int P(\alpha) \alpha^n (\alpha^*)^m e^{-|\alpha|^2} d^2\alpha. \qquad (7.12)$$

This form for the density matrix indicates a fundamental property of the fields which are most naturally described by means of the P representation. If $P(\alpha)$ is a weight function with singularities no stronger than those of delta function type, it will, in general, possess nonvanishing complex moments of arbitrarily high order. [The unique exception is the choice $P(\alpha) = \delta^{(2)}(\alpha)$ which corresponds to the ground state of the mode.] It follows then that the diagonal matrix elements $\langle n | \rho | n \rangle$, which represent the probabilities for the presence of n photons in the mode, take on nonvanishing values for arbitrarily large n. There is thus no upper bound to the number of photons present when the function P is well behaved in the sense we have noted.[15]

Stationary density operators correspond in the P representation to functions $P(\alpha)$ which depend only on $|\alpha|$. This correspondence is made clear by Eq. (7.2) which shows that such $P(\alpha)$ lead to functions $R(\beta^*,\gamma)$ which are unaltered by a common phase change of β and γ. It is seen equally well through Eq. (7.12) which shows that $\langle n | \rho | m \rangle$ reduces to diagonal form when the weight function $P(\alpha)$ is circularly symmetric.

Some indication of the importance, in practical terms, of the P representation for the density operator can be found by considering the way in which photon fields produced by different sources become superposed. Since we are only discussing the behavior of one mode of the field for the present, we are only dealing with a fragment of the full problem, but all the modes may eventually be treated similarly. We shall illustrate the superposition law by assuming there are two different transient radiation sources coupled to the field mode and that they may be switched on and off separately. The first source will be assumed, when it is turned on alone at time t_1, to excite the mode from its ground state $|0\rangle$ to the coherent state $|\alpha_1\rangle$. If we assume that the source has ceased radiating by a time t_2, the state of the field remains $|\alpha_1\rangle$ for all later times. We may alternatively consider the case in which the first source remains inactive and the second one is switched on at

---

[13] An example of a weight function $P(\alpha)$ which takes on negative values but leads to a positive-definite density operator is given by the form

$$P(\alpha) = (1+\lambda)(\pi n)^{-1} \exp[-|\alpha|^2/n] - \lambda \delta^{(2)}(\alpha)$$

for $n > 0$ and $0 < \lambda < n^{-1}$. The matrix representation of the corresponding density operator, which is given by Eq. (7.12), is seen to be diagonal and to have only positive eigenvalues.

[14] A familiar example of a function which plays a role analogous to that of a probability density, but may take on negative values in quantum-mechanical contexts is the Wigner distribution function, E. P. Wigner, Phys. Rev. **40**, 749 (1932).

[15] Density operators for fields in which the number of photons present possesses an upper bound N are represented by functions $R(\beta^*,\gamma)$ which are polynomials of Nth degree in β^* and in γ. It is evident from the behavior of such polynomials for large $|\beta|$ and $|\gamma|$ that any weight function $P(\alpha)$ which corresponds to $R(\beta^*,\gamma)$ through Eq. (7.2) would have to have singularities much stronger than those of a delta function. Such fields are probably represented more conveniently by means of the R function.

time t_2. The second source will then be assumed to bring the mode from its ground state to the coherent state $|\alpha_2\rangle$. We now ask what state the mode will be brought to if the two sources are allowed to act in succession, the first at t_1 and the second at t_2.

The answer for this simple case may be seen without performing any detailed calculations by making use of the unitary displacement operators described in Sec. III. The action of the first source is represented by the unitary operator $D(\alpha_1)$ which displaces the oscillator state from the ground state to the coherent state $|\alpha_1\rangle = D(\alpha_1)|0\rangle$. The action of the second source is evidently represented by the displacement operator $D(\alpha_2)$, so that when it is turned on after the first source, it brings the oscillator to the superposed state

$$|\ \rangle = D(\alpha_2)D(\alpha_1)|0\rangle. \qquad (7.13)$$

Since the displacement operators are of the exponential form (3.17), their multiplication law is given by Eq. (3.20). We thus find

$$D(\alpha_2)D(\alpha_1) = D(\alpha_1+\alpha_2)\exp[\tfrac{1}{2}(\alpha_2\alpha_1^* - \alpha_2^*\alpha_1)]. \quad (7.14)$$

The exponential which has been separated from the D operators in this relation has a purely imaginary argument and, hence, corresponds to a phase factor. The superposed state, (7.13), in other words, is just the coherent state $|\alpha_1+\alpha_2\rangle$ multiplied by a phase factor. The phase factor has no influence upon the density operator for the superposed state, which is

$$\rho = |\alpha_1+\alpha_2\rangle\langle\alpha_1+\alpha_2|. \qquad (7.15)$$

To vary the way in which the sources are turned on in the imaginary experiment we have described, e.g., to turn the two sources on at other times or in the reverse order, would only alter the final state through a phase factor and would thus lead to the same final density operator. The amplitudes of successive coherent excitations of the mode add as complex numbers in quantum theory, just as they do in classical theory.

Let us suppose next that the sources in the same experiment are somewhat less ideal and that, instead of exciting the mode to pure coherent states, they excite it to conditions described by mixtures of coherent states of the form (7.6). The first source acting alone, we assume, brings the field to a condition described by the density operator

$$\rho_1 = \int P_1(\alpha_1)|\alpha_1\rangle\langle\alpha_1|d^2\alpha_1. \qquad (7.16)$$

The condition produced by the second source, when it acts alone, is assumed to be represented by

$$\rho = \int P_2(\alpha_2)|\alpha_2\rangle\langle\alpha_2|d^2\alpha_2,$$

$$= \int P_2(\alpha_2)D(\alpha_2)|0\rangle\langle 0|D^{-1}(\alpha_2)d^2\alpha_2.$$

If the second source is turned on after the first, it brings the field to a condition described by the density operator

$$\rho = \int P_2(\alpha_2)D(\alpha_2)\rho_1 D^{-1}(\alpha_2)d^2\alpha_2,$$

$$= \int P_2(\alpha_2)P_1(\alpha_1)|\alpha_1+\alpha_2\rangle\langle\alpha_1+\alpha_2|d^2\alpha_1 d^2\alpha_2. \quad (7.17)$$

The latter density operator may be written in the general form

$$\rho = \int P(\alpha)|\alpha\rangle\langle\alpha|d^2\alpha,$$

if we define the weight function $P(\alpha)$ for the superposed excitations to be

$$P(\alpha) = \int \delta^{(2)}(\alpha-\alpha_1-\alpha_2)P_1(\alpha_1)P_2(\alpha_2)d^2\alpha_1 d^2\alpha_2, \quad (7.18)$$

$$= \int P_1(\alpha-\alpha')P_2(\alpha')d^2\alpha'. \qquad (7.19)$$

We see immediately from Eq. (7.18) that P is correctly normalized if P_1 and P_2 are. The simple convolution law for combining the weight functions is one of the unique features of the description of fields by means of the P representation. It is quite analogous to the law we would use in classical theory to describe the probability distribution of the sum of two uncertain Fourier amplitudes for a mode.

The convolution theorem can often be used to separate fields into component fields with simpler properties. Suppose we have a field described by a weight function $P(\alpha)$ which has a mean value of α given by

$$\bar{\alpha} = \int \alpha P(\alpha)d^2\alpha. \qquad (7.20)$$

It is clear from Eq. (7.19) that any such field may be regarded as the sum of a pure coherent field which corresponds to the weight function $\delta^{(2)}(\alpha-\bar{\alpha})$ and an additional field represented by $P(\alpha+\bar{\alpha})$ for which the mean value of α vanishes. Fields with vanishing mean values of α will be referred to as unphased fields.

The use of the P representation of the density operator, where it is not too singular, leads to simplifications in the calculation of statistical averages which go somewhat beyond those discussed in the last section. Thus, for example, the statistical average of any normally ordered product of the creation and annihilation operators, such as $(a^\dagger)^n a^m$, reduces to a simple average of $(\alpha^*)^n \alpha^m$ taken with respect to the weight function $P(\alpha)$, i.e., we have

$$\mathrm{tr}\{\rho(a^\dagger)^n a^m\} = \int P(\alpha)\langle\alpha|(a^\dagger)^n a^m|\alpha\rangle d^2\alpha,$$

$$= \int P(\alpha)(\alpha^*)^n \alpha^m d^2\alpha. \quad (7.21)$$

This identity means, in practice, that many quantum-mechanical calculations can be carried out by means which are analogous to those already familiar from classical theory.

The mean number of photons which are present in a mode is the most elementary measure of the intensity of its excitation. The operator which represents the number of photons present is seen from Eq. (2.18) to be $a^\dagger a$. The average photon number, written as $\langle n \rangle$, is therefore given by

$$\langle n \rangle = \mathrm{tr}\{\rho a^\dagger a\}. \quad (7.22)$$

According to Eq. (7.21), with its two exponents set equal to unity, we have

$$\langle n \rangle = \int P(\alpha)|\alpha|^2 d^2\alpha, \quad (7.23)$$

i.e., the average photon number is just the mean squared absolute value of the amplitude α. When two fields described by distributions P_1 and P_2 are superposed, the resulting intensities are found from rules of the form which have always been used in classical electromagnetic theory. For unphased fields the intensities add "incoherently"; for coherent states the amplitudes add "coherently."

The use of the P representation of the density operator in describing fields brings many of the results of quantum electrodynamics into forms similar to those of classical theory. While these similarities make applications of the correspondence principle particularly clear, they must not be interpreted as indicating that classical theory is any sort of adequate substitute for the quantum theory. The weight functions $P(\alpha)$ which occur in quantum theoretical applications are not accurately interpretable as probability distributions, nor are they derivable as a rule from classical treatments of the radiation sources. They depend upon Planck's constant, in general, in ways that are unfathomable by classical or semiclassical analysis.

Since a number of calculations having to do with photon statistics have been carried out in the past by essentially classical methods, it may be helpful to discuss the relation between the P representation and the classical theory a bit further. It is worth noting in particular that the definition we have given the amplitude α as an eigenvalue of the annihilation operator is an intrinsically quantum-mechanical one. If we wish to represent a given classical field amplitude for the mode as an eigenvalue, then we see from Eq. (2.20) that the appropriate value of α has a magnitude which is proportional to $\hbar^{-1/2}$. In the dimensionless terms in which α is defined, the classical description of the mode only applies to the region $|\alpha|\gg 1$ of the complex α plane, i.e., to amplitudes of oscillation which are large compared with the range of the zero-point fluctuations present in the wave packet (3.29) and (3.30). Classical theory can therefore, in principle, only furnish us with the grossest sort of information about the weight function $P(\alpha)$. When the weight function extends appreciably into the classical regions of the plane, classical theory can only be relied upon, crudely speaking, to tell us average values of the function $P(\alpha)$ over areas whose dimensions, $|\Delta\alpha|$, are of order unity or larger. From Eq. (7.10) we see that such average values will always be positive; in the classical limit they may always be interpreted as probabilities.

VIII. THE GAUSSIAN DENSITY OPERATOR

The Gaussian function is a venerable statistical distribution, familiar from countless occurrences in classical statistics. We shall indicate in this section that it has its place in quantum field theory as well, where it furnishes the natural description of the most commonly occurring type of incoherence.[1]

Let us assume that the field mode we are studying is coupled to a number of sources which are essentially similar but are statistically independent of one another in their behavior. Such sources might, in practice, simply be several hypothetical subdivisions of one large source. If we may represent the contribution of each source (numbered $j=1, \cdots N$) to the excitation of the mode by means of a weight function $p(\alpha_j)$, we may then construct the weight function $P(\alpha)$ which describes the superposed fields by means of the generalized form of the convolution theorem

$$P(\alpha) = \int \delta^{(2)}\left(\alpha - \sum_{j=1}^N \alpha_j\right) \prod_{j=1}^N p(\alpha_j) d^2\alpha_j. \quad (8.1)$$

Since the weight functions which appear in this expression are all real valued, it is sometimes convenient to think of the amplitudes α in their arguments not as complex numbers, but as two-dimensional real vectors $\boldsymbol{\alpha}$ (i.e., $\alpha_x = \mathrm{Re}\,\alpha$, $\alpha_y = \mathrm{Im}\,\alpha$). Then if λ is an arbitrary complex number represented by the vector $\boldsymbol{\lambda}$, we may use a two-dimensional scalar product for the abbreviation

$$\mathrm{Re}\,\lambda\,\mathrm{Re}\,\alpha + \mathrm{Im}\,\lambda\,\mathrm{Im}\,\alpha = \boldsymbol{\alpha}\cdot\boldsymbol{\lambda}. \quad (8.2)$$

Using this notation, we may define the two-dimensional Fourier transform of the weight function $p(\boldsymbol{\alpha})$ as

$$\xi(\boldsymbol{\lambda}) = \int \exp(i\boldsymbol{\lambda}\cdot\boldsymbol{\alpha}) p(\boldsymbol{\alpha}) d^2\alpha. \quad (8.3)$$

The superposition law (8.1) then shows that the Fourier transform of the weight function $P(\alpha)$ is given by

$$\Xi(\lambda) = \int \exp(i\lambda \cdot \alpha) P(\alpha) d^2\alpha,$$

$$= [\xi(\lambda)]^N. \tag{8.4}$$

If the individual sources are stationary ones their weight function $p(\alpha)$ depends only on $|\alpha|$. The transform $\xi(\lambda)$ may then be approximated for small values of $|\lambda|$ by

$$\xi(\lambda) = 1 - \tfrac{1}{4}\lambda^2 \int |\alpha|^2 p(\alpha) d^2\alpha,$$

$$= 1 - \tfrac{1}{4}\lambda^2 \langle |\alpha|^2 \rangle. \tag{8.5}$$

For values of $|\lambda|$ which are smaller still (i.e., $|\lambda|^2 < N^{-1/2}\langle |\alpha|^2 \rangle^{-1}$), the transform Ξ for the superposed field may be approximated by

$$\Xi(\lambda) \approx \exp\{-\tfrac{1}{4}\lambda^2 N \langle |\alpha|^2 \rangle\}. \tag{8.6}$$

Since the weight function $p(\alpha)$ may take on negative values it is necessary at this point to verify that the second moment $\langle |\alpha|^2 \rangle$ is positive. That it is indeed positive is indicated by Eqs. (7.22) and (7.23) which show that $\langle |\alpha|^2 \rangle$ is the mean number of photons which would be radiated by each source in the absence of the others. For large values of N the transform $\Xi(\lambda)$ therefore decreases rapidly as $|\lambda|$ increases. Since the function becomes vanishingly small for $|\lambda|$ lying outside the range of approximation noted earlier, we may use (8.6) more generally as an asymptotic approximation to $\Xi(\lambda)$ for large N. When we calculate the transform of this asymptotic expression for $\Xi(\lambda)$ we find

$$P(\alpha) = (2\pi)^{-2} \int \exp(-i\alpha \cdot \lambda) \Xi(\lambda) d^2\lambda,$$

$$= \frac{1}{\pi N \langle |\alpha|^2 \rangle} \exp(-\alpha^2/N\langle |\alpha|^2 \rangle). \tag{8.7}$$

The mean value of $|\alpha|^2$ for such a weight function is evidently $N\langle |\alpha|^2 \rangle$, but by the general theorem expressed in Eq. (7.23), this mean value is just the average of the total number of quanta present in the mode. If we write the latter average as $\langle n \rangle$, and resume the use of the complex notation for the variable α, the weight function (8.7) may be written as

$$P(\alpha) = \frac{1}{\pi \langle n \rangle} e^{-|\alpha|^2/\langle n \rangle}. \tag{8.8}$$

The weight function $P(\alpha)$ is positive everywhere and takes the same form as the probability distribution for the total displacement which results from a random walk in the complex plane. However, because the coherent states $|\alpha\rangle$ are not an orthogonal set, $P(\alpha)$ can only be accurately interpreted as a probability distribution for $\langle n \rangle \gg 1$. We may note that it is not ultimately necessary, in order to derive Eq. (8.8), to assume that the weight functions corresponding to the individual sources are all the same. All that is required to carry out the proof is that the moments of the individual functions be of comparable magnitudes. The mean squared value of $|\alpha|$ is then given more generally by $\sum_j \langle |\alpha_j|^2 \rangle$, rather than the value in Eq. (8.7), but this value is still the mean number of quanta in the mode, as indicated in Eq. (8.8).

It should be clear from the conditions of the derivation that the Gaussian distribution $P(\alpha)$ for the excitation of a mode possesses extremely wide applicability. The random or chaotic sort of excitation it describes is presumably characteristic of most of the familiar types of noncoherent macroscopic light sources, such as gas discharges, incandesant radiators, etc.

The Gaussian density operator

$$\rho = \frac{1}{\pi\langle n\rangle} \int e^{-|\alpha|^2/\langle n\rangle} |\alpha\rangle\langle\alpha| d^2\alpha \tag{8.9}$$

may be seen to take on a very simple form as well in the basis which specifies the photon numbers. To find this form we substitute in Eq. (8.9) the expansions (3.7) and (3.8) for the coherent states and note the identity

$$\pi^{-1}(l!m!)^{-1/2} \int \exp[-C|\alpha|^2] \alpha^l (\alpha^*)^m d^2\alpha = \delta_{lm} C^{-(m+1)},$$

which holds for $C > 0$. If we write $C = (1 + \langle n \rangle)/\langle n \rangle$ we then find

$$\rho = \frac{1}{1+\langle n\rangle} \sum_m \left\{ \frac{\langle n\rangle}{1+\langle n\rangle} \right\}^m |m\rangle\langle m|. \tag{8.10}$$

In other words, the number of quanta in the mode is distributed according to the powers of the parameter $\langle n\rangle/(1+\langle n\rangle)$. The Planck distribution for blackbody radiation furnishes an illustration of a density operator which has long been known to take the form of Eq. (8.10). The thermal excitation which leads to the blackbody distribution is an ideal example of the random type we have described earlier, and so it should not be surprising that this distribution is one of the class we have derived. It is worth noting, in particular, that while the Planck distribution is characteristic of thermal equilibrium, no such limitation is implicit in the general form of the density operator (8.9). It will apply whenever the excitation has an appropriately random quality, no matter how far the radiator is from thermal equilibrium.

The Gaussian distribution function $\exp[-|\alpha|^2/\langle n\rangle]$ is phrased in terms which are explicitly quantum mechanical. In the limit which would represent a classical field both $|\alpha|^2$ and the average quantum number $\langle n \rangle$ become infinite as $\hbar^{-1}$, but their quotient, which is the argument of the Gaussian function, remains well defined. The form which the distribution takes in the classical limit is a familiar one. Historically, one of the origins of the random walk problem is to be found in the discussion of a classical harmonic oscillator which is subject to random excitations.[16] Such oscillators have complex amplitudes which are described under quite general conditions by a Gaussian distribution. If we were armed with this knowledge, and lacked the quantum-mechanical analysis given earlier, we might be tempted to assume that a Gaussian distribution derived in this way from classical theory can describe the photon distribution. To demonstrate the fallacy of this view we must examine more closely the nature of the parameter $\langle n \rangle$ which is, after all, the only physical constant involved in the distribution. We may take, as a simple illustration, the case of thermal excitation corresponding to temperature T. Then the mean photon number is given by $\langle n \rangle = [\exp(\hbar\omega/\kappa T) - 1]^{-1}$, where κ is Boltzmann's constant, and the distribution $P(\alpha)$ takes the form

$$P(\alpha) = \frac{1}{\pi}[e^{\hbar\omega/\kappa T} - 1] \exp[-(e^{\hbar\omega/\kappa T} - 1)|\alpha|^2]. \quad (8.11)$$

To reach the classical analog of this distribution we would assume that the classical field energy in the mode, $H = \frac{1}{2}\int(\mathbf{E}^2 + \mathbf{B}^2)d\mathbf{r}$, is distributed with a probability proportional to $\exp[-H/\kappa T]$. The distribution for the amplitude α that results is

$$P_{cl}(\alpha) = (\hbar\omega/\pi\kappa T) \exp[-\hbar\omega|\alpha|^2/\kappa T], \quad (8.12)$$

which is seen to be a first approximation in powers of $\hbar$ to the correct distribution. (Again, we must remember that the quantity $\hbar|\alpha|^2$ is to be construed as a classical parameter.) The distribution $P_{cl}(\alpha)$ only extends into the classical region of the plane, $|\alpha| \gg 1$, for low-frequency modes, that is, only for $(\hbar\omega/\kappa T) \ll 1$ are the modes sufficiently excited to be accurately described by classical theory. For higher frequencies the two distributions differ greatly in nature even though both are Gaussian. The classical distribution retains much too large a radius in the α plane as $\hbar\omega$ increases beyond κT, rather than narrowing extremely rapidly as the correct distribution does.[17] That error, in fact, epitomizes the ultraviolet catastrophe of the classical radiation theory. The example we have discussed is, of course, an elementary one, but it should serve to illustrate some of the points noted in the preceding section regarding the limitations of the classical distribution function.

The expression for the thermal density operator of an oscillator in terms of coherent quantum states appears to offer new and instructive approaches to many familiar problems. It permits us, for example, to derive the thermal averages of exponential functions of the operators a and $a^\dagger$ in an elementary way. The thermal average of the operator $D(\beta)$ defined by Eq. (3.17) is an illustration. It is given by

$$\mathrm{tr}\{\rho D(\beta)\} = \frac{1}{\pi\langle n \rangle} \int e^{-|\alpha|^2/\langle n \rangle}\langle \alpha|D(\beta)|\alpha\rangle d^2\alpha. \quad (8.13)$$

The expectation value in the integrand is, in this case

$$\begin{aligned}\langle\alpha|D(\beta)|\alpha\rangle &= \langle 0|D^{-1}(\alpha)D(\beta)D(\alpha)|0\rangle, \\ &= \exp[\beta\alpha^* - \beta^*\alpha]\langle 0|D(\beta)|0\rangle, \\ &= \exp[\beta\alpha^* - \beta^*\alpha]\langle 0|\beta\rangle, \\ &= \exp[\beta\alpha^* - \beta^*\alpha - \tfrac{1}{2}|\beta|^2], \quad (8.14)\end{aligned}$$

where the properties of $D(\alpha)$ as a displacement operator have been used in the intermediate steps. When the integration indicated in Eq. (8.13) is carried out, we find

$$\mathrm{tr}\{\rho D(\beta)\} = \exp[-|\beta|^2(\langle n \rangle + \tfrac{1}{2})], \quad (8.15)$$

which is a frequently used corollary of Bloch's theorem on the distribution function of an oscillator coordinate.[18]

IX. DENSITY OPERATORS FOR THE FIELD

The developments introduced in Secs. III–VIII have all concerned the description of the quantum state of a single mode of the electromagnetic field. We may describe the field as a whole by constructing analogous methods to deal with all its modes at once. For this purpose we introduce a basic set of coherent states for the entire field and write them as

$$|\{\alpha_k\}\rangle \equiv \prod_k |\alpha_k\rangle_k, \quad (9.1)$$

where the notation $\{\alpha_k\}$, which will be used in several other connections, stands for the set of all amplitudes α_k. It is clear then, from the arguments of Sec. IV, that any state of the field determines uniquely a function $f(\{\alpha_k^*\})$ which is an entire function of each of the variables α_k^*. If the Hilbert space vector which represents the state is known and designated as $|f\rangle$, the function f is given by

$$f(\{\alpha_k^*\}) = \langle\{\alpha_k\}|f\rangle \exp(\tfrac{1}{2}\sum_k|\alpha_k|^2), \quad (9.2)$$

which is the direct generalization of Eq. (4.11). The expansion for the state $|f\rangle$ in terms of coherent states is then

$$|f\rangle = \int |\{\alpha_k\}\rangle f(\{\alpha_k^*\}) \prod_k \pi^{-1} e^{-\frac{1}{2}|\alpha_k|^2} d^2\alpha_k, \quad (9.3)$$

which generalizes Eq. (4.7).

All of the operators which occur in field theory possess expansions in terms of the vectors $|\{\alpha_k\}\rangle$ and their

---

[16] Lord Rayleigh, *The Theory of Sound*, (MacMillan and Company Ltd., London, 1894), 2nd ed., Vol. I, p. 35; *Scientific Papers* (Cambridge University Press, Cambridge, England, 1899–1920), Vol. I, p. 491, Vol. IV, p. 370.

[17] For frequencies in the middle of the visible spectrum and temperatures under 3000°K the quantum mechanical distribution (8.11) will have a radius which corresponds to $|\alpha|^2 \ll 10^{-3}$, i.e., the distribution is far from classical in nature. Comparable radii characterize the distributions for nonthermal incoherent sources.

[18] F. Bloch, Z. Physik 74, 295 (1932).

adjoints. To construct such representations is simply a matter of generalizing the formulas of Sec. V to deal with an infinite set of amplitude variables. We therefore proceed directly to a discussion of the density operator. For any density operator ρ we may define a function $R(\{\alpha_k^*\},\{\beta_k\})$ which is an entire function of each of the variables α_k^* and β_k for all modes k. This function, as may be seen from Eq. (6.1), is given by

$$R(\{\alpha_k^*\},\{\beta_k\}) = \langle\{\alpha_k\}|\rho|\{\beta_k\}\rangle$$
$$\times \exp[\tfrac{1}{2}\sum (|\alpha_k|^2+|\beta_k|^2)]. \quad (9.4)$$

The corresponding representation of the density operator is

$$\rho = \int |\{\alpha_k\}\rangle R(\{\alpha_k^*\},\{\beta_k\})\langle\{\beta_k\}|\prod_k \pi^{-2}$$
$$\times e^{-\tfrac{1}{2}(|\alpha_k|^2+|\beta_k|^2)}d^2\alpha_k d^2\beta_k. \quad (9.5)$$

If the set of integers $\{n_k\}$ is used to specify the familiar stationary states which have n_k photons in the kth mode, we may regard R as a generating function for the matrix elements of ρ connecting these states, i.e., as a generalization of Eq. (6.2) we have

$$R(\{\alpha_k^*\},\{\beta_k\}) = \sum_{\{n_k\},\{m_k\}} \langle\{n_k\}|\rho|\{m_k\}\rangle$$
$$\times \prod_k (n_k!m_k!)^{-1/2}(\alpha_k^*)^{n_k}\beta_k^{m_k}. \quad (9.6)$$

The matrix elements of ρ in the stationary basis are then given by

$$\langle\{n_k\}|\rho|\{m_k\}\rangle$$
$$= \int R(\{\alpha_k^*\},\{\beta_k\})\prod_k \pi^{-2}(n_k!m_k!)^{-1/2}\alpha_k^{n_k}(\beta_k^*)^{m_k}$$
$$\times e^{-(|\alpha_k|^2+|\beta_k|^2)}d^2\alpha_k d^2\beta_k. \quad (9.7)$$

The normalization condition on R is clearly

$$\int R(\{\beta_k^*\},\{\beta_k\})\prod_k \pi^{-1}e^{-|\beta_k|^2}d^2\beta_k = 1. \quad (9.8)$$

The positive definiteness condition, Eq. (6.9), may also be generalized in an evident way to deal with the full set of amplitude variables.

It may help as a simple illustration of the foregoing formulae to consider the representation of a single-photon wave packet. The state which is empty of all photons is the one for which the amplitudes α_k all vanish. If we write that state as $|\text{vac}\rangle$, then we may write the most general one-photon state as $\sum_k q(k)a_k^\dagger|\text{vac}\rangle$, where the function $q(k)$ plays the role of a packet amplitude. The function f which represents this state is then

$$f(\{\alpha_k^*\}) = \sum_k q(k)\alpha_k^*, \quad (9.9)$$

and the corresponding function R which determines the density operator is

$$R(\{\alpha_k^*\},\{\beta_k\}) = \sum_k q(k)\alpha_k^* \sum_{k'} q^*(k')\beta_{k'}. \quad (9.10)$$

The normalization condition (9.8) corresponds to the requirement $\sum |q(k)|^2 = 1$. Since the state we have considered is a pure one, the function R factorizes into the product of two functions, one having the form of f and the other of its complex conjugate. If the packet amplitudes $q(k)$ were in some degree unpredictable, as they usually are, the packet could no longer be represented by a pure state. The function R would then be an average taken over the distribution of the amplitudes $q(k)$ and hence would lose its factorizable form in general. Whenever an upper bound exists for the number of photons present, i.e., the number of photons is required to be less than or equal to some integer N, we will find that R is a polynomial of at most Nth degree in the variables $\{\alpha_k^*\}$ and of the same degree in the $\{\beta_k\}$.

There will, of course, exist many types of excitation for which the photon numbers are unbounded. Among these are the ones which are more conveniently described by means of a generalized P distribution, i.e., the excitations for which there exists a reasonably well-behaved real-valued function $P(\{\alpha_k\})$ such that

$$R(\{\beta_k^*\},\{\gamma_k\}) = \int P(\{\alpha_k\})$$
$$\times \exp\left[\sum_k (\beta_k^*\alpha_k+\gamma_k\alpha_k^*-|\alpha_k|^2)\right]\prod_k d^2\alpha_k. \quad (9.11)$$

When R possesses a representation of this type the density operator (9.5) may be reduced by means of Eq. (4.14) and its complex conjugate to the simple form

$$\rho = \int P(\{\alpha_k\})|\{\alpha_k\}\rangle\langle\{\alpha_k\}|\prod_k d^2\alpha_k, \quad (9.12)$$

which is the many-mode form of the P representation given by Eq. (7.6). The function P must satisfy the positive definiteness condition

$$\int |f(\{\alpha_k^*\})|^2 P(\{\alpha_k\})\prod_k e^{-|\alpha_k|^2}d^2\alpha_k \geq 0 \quad (9.13)$$

for all possible choices of entire functions $f(\{\alpha_k^*\})$. The matrix elements of the density operator in the representation based on the n-photon states are

$$\langle\{n_k\}|\rho|\{m_k\}\rangle = \int P(\{\alpha_k\})$$
$$\times \prod_k (n_k!m_k!)^{-1/2}\alpha_k^{n_k}(\alpha_k^*)^{m_k}e^{-|\alpha_k|^2}d^2\alpha_k. \quad (9.14)$$

Stationary density operators, i.e., ones which commute with the Hamiltonian correspond to functions $P(\{\alpha_k\})$ which depend on the amplitude variables only through their magnitudes $\{|\alpha_k|\}$.

The superposition of two fields is described by forming the convolution integral of their distribution functions, much as in the case of a single mode. Thus, if two fields, described by $P_1(\{\beta_k\})$ and $P_2(\{\gamma_k\})$, respectively, are superposed, the resulting field has a distribution function

$$P(\{\alpha_k\}) = \int \prod_k \delta^{(2)}(\alpha_k - \beta_k - \gamma_k)$$
$$\times P_1(\{\beta_k\}) P_2(\{\gamma_k\}) \prod_k d^2\beta_k d^2\gamma_k. \quad (9.15)$$

For fields which are represented by means of the density operator (9.12) all of the averages of normally ordered operator products can be calculated by means of formulas which, as in the case of a single mode, greatly resemble those of classical theory. Thus, the parameters $\{\alpha_k\}$ play much the same role in these calculations as the random Fourier amplitudes of the field do in the familiar classical theory of microwave noise.[19] Furthermore, the weight function $P(\{\alpha_k\})$ plays a role similar to that of the probability distribution for the Fourier amplitudes. Although this resemblance is extremely convenient in calculations, and offers immediate insight into the application of the correspondence principle, we must not lose sight of the fact that the function $P(\{\alpha_k\})$ is, in general, an explicitly quantum-mechanical structure. It may assume negative values, and is not accurately interpretable as a probability distribution except in the classical limit of strongly excited or low frequency fields.

In the foregoing discussions we have freely assumed that the density operator which describes the field is known and that it may, therefore, be expressed either in the representation of Eq. (9.5) or in the P representation of Eq. (9.12). For certain types of incoherent sources which we have discussed in Sec. VIII and will mention again in Sec. X, the explicit construction of these density operators is not at all difficult. But to find accurate density operators for other types of sources, including the recently developed coherent ones, will require a good deal of physical insight. The general problem of treating quantum mechanically the interaction of a many-atom source both with the radiation field and with an excitation mechanism of some sort promises to be a complicated one. It will have to be approached, no doubt, through greatly simplified models.

Since very little is known about the density operator for radiation fields, some insight may be gained by examining the form it takes on in one of the few completely soluble problems of quantum electrodynamics. We shall study the photon field radiated by an electric current distribution which is essentially classical in nature, one that does not suffer any noticeable reaction from the process of radiation. We may then represent the radiating current by a prescribed vector function of space and time $\mathbf{j}(\mathbf{r},t)$. The Hamiltonian which describes the coupling of the quantized electromagnetic field to the current distribution takes the form

$$H_1(t) = -\frac{1}{c} \int \mathbf{j}(\mathbf{r},t) \cdot \mathbf{A}(\mathbf{r},t) d\mathbf{r}. \quad (9.16)$$

The introduction of an explicitly time-dependent interaction of this type means that the state vector for the field, $|\ \rangle$, which previously was fixed (corresponding to the Heisenberg picture) will begin to change with time in accordance with the Schrödinger equation

$$i\hbar \frac{\partial}{\partial t} |\ \rangle = H_1(t) |\ \rangle, \quad (9.17)$$

which is the one appropriate to the interaction representation. The solution of this equation is easily found.[20] If we assume that the initial state of the field at time $t = -\infty$ is one empty of all photons, then the state of the field at time t may be written in the form

$$|t\rangle = \exp\left\{\frac{i}{\hbar c} \int_{-\infty}^{t} dt' \int \mathbf{j}(\mathbf{r},t') \cdot \mathbf{A}(\mathbf{r},t') d\mathbf{r} + i\varphi(t)\right\} |\text{vac}\rangle.$$
$$(9.18)$$

The function $\varphi(t)$ which occurs in the exponent is a real-valued c-number phase function. It is easily evaluated, but cancels out of the product $|t\rangle\langle t|$ and so has no bearing on the construction of the density operator. The exponential operator which occurs in Eq. (9.18) may be expressed quite simply in terms of the displacement operators we discussed in Sec. III. For this purpose we define a displacement operator D_k for the kth mode as

$$D_k(\beta_k) = \exp[\beta_k \alpha_k^\dagger - \beta_k^* \alpha_k]. \quad (9.19)$$

Then it is clear from the expansion (2.10) for the vector potential that we may write

$$\exp\left\{\frac{i}{\hbar c} \int_{-\infty}^{t} dt' \int \mathbf{j}(\mathbf{r},t') \cdot \mathbf{A}(\mathbf{r},t') d\mathbf{r}\right\} = \prod_k D_k[\alpha_k(t)], \quad (9.20)$$

where the time-dependent amplitudes $\alpha_k(t)$ are given by

$$\alpha_k(t) = \frac{i}{(2\hbar\omega)^{1/2}} \int_{-\infty}^{t} dt' \int d\mathbf{r}\, \mathbf{u}_k^*(\mathbf{r}) \cdot \mathbf{j}(\mathbf{r},t') e^{i\omega t'}. \quad (9.21)$$

The density operator at time t may therefore be written

---

[19] J. Lawson and G. E. Uhlenbeck, *Threshold Noise Signals* (McGraw-Hill Book Company, Inc., New York, 1950), pp. 33–56.

[20] R. J. Glauber, Phys. Rev. **84**, 395 (1951).

as

$$|t\rangle\langle t| = \prod_k D_k[\alpha_k(t)] |\text{vac}\rangle\langle\text{vac}| \prod_k D_k^{-1}[\alpha_k(t)] \quad (9.22)$$

$$= |\{\alpha_k(t)\}\rangle\langle\{\alpha_k(t)\}| . \quad (9.23)$$

The radiation by any prescribed current distribution, in other words, always leads to a pure coherent state.

It is only a slight generalization of the model we have just considered to imagine that the current distribution $\mathbf{j}(\mathbf{r},t)$ is not wholly predictable. In that case the amplitudes $\alpha_k(t)$ defined by Eq. (9.21) become random variables which possess collectively a probability distribution function which we may write as $p(\{\alpha_k\},t)$. The density operator for the field radiated by such a random current then becomes

$$\rho(t) = \int p(\{\alpha_k\},t) |\{\alpha_k\}\rangle\langle\{\alpha_k\}| \prod_k d^2\alpha_k. \quad (9.24)$$

We see that the density operator for a field radiated by a random current which suffers no recoil in the radiation process always takes the form of the P representation of Eq. (9.12). The weight function in this case does admit interpretation as a probability distribution, but it has a classical structure associated directly with the properties of the radiating current rather than with particular (nonorthogonal) states of the field. The assumption we have made in defining the model, that the current suffers negligible reaction, is a strong one but is fairly well fulfilled in radiating systems operated at radio or microwave frequencies. The fields produced by such systems should be accurately described by density operators of the form (9.24).

X. CORRELATION AND COHERENCE PROPERTIES OF THE FIELD

Any eigenvalue function $\mathcal{E}(\mathbf{r}t)$ which satisfies the appropriate field equations and contains only positive frequency terms determines a set of mode amplitudes $\{\alpha_k\}$ uniquely through the expansion (2.20). This set of mode amplitudes then determines a coherent state of the field, $|\{\alpha_k\}\rangle$, such that

$$\mathbf{E}^{(+)}(\mathbf{r}t)|\{\alpha_k\}\rangle = \mathcal{E}(\mathbf{r}t)|\{\alpha_k\}\rangle. \quad (10.1)$$

To discuss the general form which the field correlation functions take in such states it is convenient to abbreviate a set of coordinates $(\mathbf{r}_j,t_j)$ by a single symbol x_j. The nth-order correlation function is then defined as[3]

$$G_{\mu_1\cdots\mu_{2n}}{}^{(n)}(x_1\cdots x_{2n}) = \text{tr}\{\rho E_{\mu_1}{}^{(-)}(x_1)\cdots$$
$$\times E_{\mu_n}{}^{(-)}(x_n) E_{\mu_{n+1}}{}^{(+)}(x_{n+1})\cdots E_{\mu_{2n}}{}^{(+)}(x_{2n})\}. \quad (10.2)$$

The density operator for the coherent state defined by Eq. (10.1) is the projection operator

$$\rho = |\{\alpha_k\}\rangle\langle\{\alpha_k\}| . \quad (10.3)$$

For this operator it follows from Eq. (10.1) and its Hermitian adjoint that the correlation functions reduce to the factorized form

$$G_{\mu_1\cdots\mu_{2n}}{}^{(n)}(x_1\cdots x_{2n}) = \prod_{j=1}^{n} \mathcal{E}_{\mu_j}{}^*(x_j) \prod_{l=n+1}^{2n} \mathcal{E}_{\mu_l}(x_l). \quad (10.4)$$

In other words, the field which corresponds to the state $|\{\alpha_k\}\rangle$ satisfies the conditions for full coherence according to the definition[3] given earlier.

It is worth noting that the state $|\{\alpha_k\}\rangle$ is not the only one which leads to the set of correlation functions (10.4). Indeed, let us consider a state which corresponds not to the amplitudes $\{\alpha_k\}$, but to a set $\{e^{i\varphi}\alpha_k\}$ which differs by a common phase factor (i.e., φ is real and independent of k). Then the corresponding eigenvalue function becomes $e^{i\varphi}\mathcal{E}(\mathbf{r}t)$, but such a change leaves the correlation functions (10.4) unaltered. It is clear from this invariance property of the correlation functions that certain mixtures of the coherent states also lead to the same set of functions. Thus, if $|\{\alpha_k\}\rangle$ is the state defined by Eq. (10.1), and $\mathcal{L}(\varphi)$ is any real-valued function of φ normalized in the sense

$$\int_0^{2\pi} \mathcal{L}(\varphi) d\varphi = 1, \quad (10.5)$$

we see that the density operator

$$\rho = \int_0^{2\pi} \mathcal{L}(\varphi) |\{e^{i\varphi}\alpha_k\}\rangle\langle\{e^{i\varphi}\alpha_k\}| d\varphi \quad (10.6)$$

leads for all choices of $\mathcal{L}(\varphi)$ to the set of correlation functions (10.4). Such a density operator is, of course, a special case of the general form (9.12), one which corresponds to an over-all uncertainty in the phase of the $\{\alpha_k\}$. The particular choice $\mathcal{L}(\varphi) = (2\pi)^{-1}$, which corresponds to complete ignorance of the phase, represents the usual state of our knowledge about high-frequency fields. We have been careful, therefore, to define coherence in terms of a set of correlation functions which are independent of the over-all phase.

Since nonstationary fields of many sorts can be represented by means of eigenvalue functions, it becomes a simple matter to construct corresponding quantum states. As an illustration we may consider the example of an amplitude-modulated plane wave. For this purpose we make use of the particular set of mode functions defined by Eq. (2.9). Then if the carrier wave has frequency ω and the modulation is periodic and has frequency $\zeta\omega$ where $0 < \zeta < 1$, we may write an appropriate eigenvalue function as

$$\mathcal{E}(\mathbf{r}t) = i\left(\frac{\hbar\omega}{2L^3}\right)^{1/2} \hat{e}^{(\lambda)}\alpha_\mathbf{k}$$
$$\times \{1 + M\cos[\zeta(\mathbf{k}\cdot\mathbf{r} - \omega t) - \delta]\} e^{i(\mathbf{k}\cdot\mathbf{r} - \omega t)}. \quad (10.7)$$

When this expression is expanded in plane-wave modes it has only three nonvanishing amplitude coefficients. These are α_k itself and the two sideband amplitudes

$$\alpha_{k(1-\zeta)} = \tfrac{1}{2}M(1-\zeta)^{-1/2}e^{i\delta}\alpha_k,$$
$$\alpha_{k(1+\zeta)} = \tfrac{1}{2}M(1+\zeta)^{-1/2}e^{-i\delta}\alpha_k. \qquad (10.8)$$

The coherent state which corresponds to the modulated wave may be constructed immediately from the knowledge of these amplitudes. In practice, of course, we will not often know the phase of α_k, and so the wave should be represented not by a single coherent state, but by a mixture of the form (10.6). Representations of other forms of modulated waves may be constructed similarly.

Incoherent fields, or the broad class of fields for which the correlation functions do not factorize, must be described by means of density operators which are more general in their structure than those of Eqs. (10.3) or (10.6). To illustrate the form taken by the correlation functions for such cases we may suppose the field to be described by the P representation of the density operator. Then the first-order correlation function is given by

$$G_{\mu\nu}^{(1)}(\mathbf{r}t,\mathbf{r}'t') = \int P(\{\alpha_k\}) \sum_{k,k'} \tfrac{1}{2}\hbar(\omega\omega')^{1/2} u_{k\mu}^*(\mathbf{r}) u_{k'\nu}(\mathbf{r}')$$
$$\times \alpha_k^* \alpha_{k'} e^{i(\omega t - \omega' t')} \prod_l d^2\alpha_l. \qquad (10.9)$$

Fields for which the P representation is inconveniently singular may, as we have noted earlier, always be described by means of analytic functions $R(\{\alpha_k^*\},\{\beta_k\})$ and corresponding density operators of the form (9.5). When that form of density operator is used to evaluate the first-order correlation function we find

$$G_{\mu\nu}^{(1)}(\mathbf{r}t,\mathbf{r}'t') = \int R(\{\alpha_k^*\},\{\beta_k\}) \sum_{k',k''} \tfrac{1}{2}\hbar(\omega'\omega'')^{1/2}$$
$$\times u_{k'\mu}^*(\mathbf{r}) u_{k''\nu}(\mathbf{r}')\beta_{k'}^*\alpha_{k''}e^{i(\omega't-\omega''t')}$$
$$\times \prod_l e^{\beta_l^*\alpha_l} d\mu(\alpha_l) d\mu(\beta_l), \qquad (10.10)$$

where the differentials $d\mu(\alpha_l)$ and $d\mu(\beta_l)$ are those defined by Eq. (5.14). The higher order correlation functions are given by integrals analogous to (10.9) and (10.10). Their integrands contain polynomials of the $2n$th degree in the amplitude variables α_k and β_k^* in place of the quadratic forms which occur in the first-order functions.

The energy spectrum of a radiation field is easily derived from a knowledge of its first-order correlation function. If we return for a moment to the expansion (2.19) for the positive-frequency field operator, and write the negative-frequency field as its Hermitian adjoint, we see that these operators obey the identity

$$2\int \mathbf{E}^{(-)}(\mathbf{r}t)\cdot\mathbf{E}^{(+)}(\mathbf{r}t')d\mathbf{r}$$
$$= \sum_k \hbar\omega a_k^\dagger a_k \exp[i\omega(t-t')]. \qquad (10.11)$$

If we take the statistical average of both sides of this equation we may write the resulting relation as

$$\sum_\mu \int G_{\mu\mu}^{(1)}(\mathbf{r}t,\mathbf{r}t')d\mathbf{r} = \tfrac{1}{2}\sum_k \hbar\omega\langle n_k\rangle \exp[i\omega(t-t')], \quad (10.12)$$

where $\langle n_k\rangle$ is the average number of photons in the kth mode. The Fourier representation of the volume integral of $\sum_\mu G_{\mu\mu}^{(1)}$ therefore identifies the energy spectrum $\hbar\omega\langle n_k\rangle$ quite generally.

For fields which may be represented by stationary density operators, it becomes still simpler to extract the energy spectrum from the correlation function. For such fields the weight function $P(\{\alpha_k\})$ depends only on the absolute values of the α_k, so that we have

$$\int P(\{\alpha_k\})\alpha_{k'}^*\alpha_{k''} \prod_l d^2\alpha_l = \langle|\alpha_{k'}|^2\rangle \delta_{k'k''}$$
$$= \langle n_{k'}\rangle \delta_{k'k''}. \qquad (10.13)$$

By using Eq. (10.9) to evaluate the correlation function, and specializing to the case of plane-wave modes, we then find

$$\sum_\mu G_{\mu\mu}^{(1)}(\mathbf{r}t,\mathbf{r}t') = \tfrac{1}{2}L^{-3}\sum_{k,\lambda}\hbar\omega\langle n_{k,\lambda}\rangle e^{i\omega(t-t')}, \quad (10.14)$$

in which we have explicitly indicated the role of the polarization index λ. If the volume which contains the field is sufficiently large in comparison to the wavelengths of the excited modes, the sum over the modes in Eq. (10.14) may be expressed as an integral over $\mathbf{k}$ space $[\sum_k \to \int L^3(2\pi)^{-3}d\mathbf{k}]$. By defining an energy spectrum for the quanta present (i.e., an energy per unit interval of ω) as

$$w(\omega) = (2\pi)^{-3}\hbar k^3 \sum_\lambda \int \langle n_{k,\lambda}\rangle d\Omega_k, \qquad (10.15)$$

where $d\Omega_k$ is an element of solid angle in $\mathbf{k}$ space, we may then rewrite Eq. (10.14) in the form

$$\sum_\mu G_{\mu\mu}^{(1)}(\mathbf{r}t,\mathbf{r}t') = \tfrac{1}{2}\int_0^\infty w(\omega)e^{i\omega(t-t')}d\omega. \quad (10.16)$$

With the understanding that $w(\omega)=0$ for $\omega<0$, we may extend the integral over ω from $-\infty$ to ∞. It is then clear that the relation (10.16) may be inverted to express the energy spectrum as the Fourier transform of the time-dependent correlation function,

$$w(\omega) = \frac{1}{\pi} \int_{-\infty}^{\infty} \sum_{\mu} G_{\mu\mu}{}^{(1)}(\mathbf{r}0, \mathbf{r}t) e^{i\omega t} dt. \quad (10.17)$$

A pair of relations analogous to Eqs. (10.16) and (10.17), and together called the Wiener-Khintchine theorem, has long been of use in the classical theory of random fields.[21] The relations we have derived are, in a sense, the natural quantum mechanical generalization of the Wiener-Khintchine theorem. All we have assumed is that the field is describable by a stationary form of the P representation of the density operator. The proof need not, in fact, rest upon the use of the P representation since we can construct a corresponding statement in terms of the more general representation (9.5).

Stationary fields, according to Eq. (6.10), are represented by entire functions $R = \mathcal{S}(\{\alpha_k{}^*\beta_k\})$, i.e., functions which depend only on the set of products $\alpha_k{}^*\beta_k$. For such fields, then, the integral over the α and β planes which is required in Eq. (10.10) takes the form

$$\langle \beta_{k'}{}^* \alpha_{k''} \rangle = \int \mathcal{S}(\{\alpha_k{}^*\beta_k\}) \beta_{k'}{}^* \alpha_{k''} \prod_l e^{\beta_l{}^* \alpha_l} d\mu(\alpha_l) d\mu(\beta_l). \quad (10.18)$$

Since the range of integration of each of the α and β variables covers the entire complex plane, this integral cannot be altered if we change the signs of any of the variables. If, however, we replace the particular variables $\alpha_{k''}$ and $\beta_{k''}$ by $-\alpha_{k''}$ and $-\beta_{k''}$ the integral is seen to reverse in sign, unless we have

$$\langle \beta_{k'}{}^* \alpha_{k''} \rangle = \delta_{k'k''} \langle \beta_{k'}{}^* \alpha_{k'} \rangle. \quad (10.19)$$

The average $\langle \beta_k{}^* \alpha_k \rangle$, we may note from Eqs. (5.11) and (6.5), is just the mean number of quanta in the kth mode,

$$\langle \beta_k{}^* \alpha_k \rangle = \text{tr}\{\rho a^\dagger_k a_k\} = \langle n_k \rangle. \quad (10.20)$$

We have thus shown that the general expression (10.10) for the first-order correlation function always satisfies Eq. (10.14) when the field is described by a stationary density operator. The derivation of the equations relating the energy spectrum to the time-dependent correlation function then proceeds as before.

The simplest and most universal example of an incoherent field is the type generated by superposing the outputs of stationary sources. We have shown in some detail in Sec. VIII that as the number of sources which contribute to the excitation of a single mode increases, the density operator for the mode takes on a Gaussian form in the P representation. It is not difficult to derive an analogous result for the case of sources

which excite many modes at once. We shall suppose that the sources ($j = 1 \cdots N$) are essentially identical, and that their contributions to the excitation are described by a weight function $p(\{\alpha_{jk}\})$. The weight function $P(\{\alpha_k\})$ for the superposed fields is then given by the convolution theorem as

$$P(\{\alpha_k\}) = \int \prod_k \delta^{(2)}\left(\alpha_k - \sum_{j=1}^{N} \alpha_{jk}\right) \prod_{j=1}^{N} p(\{\alpha_{jk}\}) \prod_k d^2\alpha_{jk}. \quad (10.21)$$

Since the individual sources are assumed to be stationary, the function $p(\{\alpha_{jk}\})$ will only depend on the variables α_{jk} through their absolute magnitudes, $|\alpha_{jk}|$.

The derivation which leads from Eq. (10.21) to a Gaussian asymptotic form for $P(\{\alpha_k\})$ is so closely parallel to that of Eqs. (8.1)–(8.8) that there is no need to write it out in detail. The argument makes use of second-order moments of the function p which may, with the same type of vector notation used previously, be written as

$$\langle \alpha_k \alpha_{k'} \rangle = \int \alpha_k \alpha_{k'} p(\{\alpha_l\}) \prod_l d^2\alpha_l. \quad (10.22)$$

The stationary character of the function p implies that such moments vanish for $k \neq k'$. With this observation, we may retrace our earlier steps to show that the many-dimensional Fourier transform of P takes the form of a product of Gaussians, one for each mode and each similar in form to that of Eq. (8.6). It then follows immediately that the weight function P for the field as a whole is given by a product of Gaussian factors each of the form of Eq. (8.8). We thus have

$$P(\{\alpha_k\}) = \prod_k \frac{1}{\pi \langle n_k \rangle} e^{-|\alpha_k|^2/\langle n_k \rangle}, \quad (10.23)$$

where $\langle n_k \rangle$ is the average number of photons present in the kth mode when the fields are fully superposed. One of the striking features of this weight function is its factorized form. It is interesting to remember, therefore, that no assumption of factorizability has been made regarding the weight functions p which describe the individual sources. These sources may, indeed, be ones for which the various mode amplitudes are strongly coupled in magnitude. It is the stationary property of the sources which leads, because of the vanishing of the moments (10.22) for $k \neq k'$, to the factorized form for the weight function (10.23).

The density operator which corresponds to the Gaussian weight function (10.23) evidently describes an ideally random sort of excitation of the field modes. We may reasonably surmise that it applies, at least as a good approximation, to all of the familiar sorts of incoherent sources in laboratory use. It is clear, in particular, from the arguments of Sec. VII that the Gaussian weight function describes thermal sources

---

[21] The Wiener-Khintchine theorem is usually expressed in terms of cosine transforms since it deals with a real-valued correlation function for the classical field $\mathbf{E}$, rather than a complex one for the fields $\mathbf{E}^{(\pm)}$. The complex correlation functions are considerably more convenient to use for quantum mechanical purposes, as is shown in Ref. 3.

correctly. The substitution of the Planck distribution $\langle n_k \rangle = [\exp(\hbar\omega_k/\kappa T) - 1]^{-1}$ into Eq. (10.23) leads to the density operator for the entire thermal radiation field. To the extent that the Gaussian weight function (10.23) may describe radiation by a great variety of incoherent sources there will be certain deep-seated similarities in the photon fields generated by all of them. One may, for example, think of these sources all as resembling thermal ones and differing from them only in the spectral distributions of their outputs. As a way of illustrating these similarities we might imagine passing blackbody radiation through a filter which is designed to give the spectral distribution of the emerging light a particular line profile. We may choose this artificial line profile to be the same as that of some true emission line radiated, say, by a discharge tube. We then ask whether measurements carried out upon the photon field can distinguish the true emission-line source from the artificial one. If the radiation by the discharge tube is described, as we presume, by a Gaussian weight function, it is clear that the two sources will be indistinguishable from the standpoint of any photon counting experiments. They are equivalent sorts of narrow-band, quantum-mechanical noise generators.

It is a simple matter to find the correlation functions for the incoherent fields[2] described by the Gaussian weight function (10.23). If we substitute this weight function into the expansion (10.9) for the first-order correlation function we find

$$G_{\mu\nu}{}^{(1)}(\mathbf{r}t,\mathbf{r}'t') = \tfrac{1}{2}\sum_{k} \hbar\omega u_{k\mu}{}^*(\mathbf{r}) u_{k\nu}(\mathbf{r}') \langle n_k \rangle e^{i\omega(t-t')}. \quad (10.24)$$

When the mode functions $\mathbf{u}_k(\mathbf{r})$ are the plane waves of Eq. (2.9), and the volume of the system is sufficiently large, we may write the correlation function as the integral

$$G_{\mu\nu}{}^{(1)}(\mathbf{r}t,\mathbf{r}'t') = \frac{\hbar c}{2(2\pi)^3} \int \sum_{\lambda} e_\mu{}^{(\lambda)*} e_\nu{}^{(\lambda)} \langle n_{\mathbf{k},\lambda} \rangle k$$

$$\times \exp\{-i[\mathbf{k}\cdot(\mathbf{r}-\mathbf{r}') - \omega(t-t')]\} d\mathbf{k}, \quad (10.25)$$

in which the index λ again labels polarizations. To find the second-order correlation function defined by Eq. (10.2) we may write it likewise as an expansion in terms of mode functions. The only new moments of the weight function which we need to know are those given by $\langle |\alpha_k|^4 \rangle = 2\langle |\alpha_k|^2 \rangle^2 = 2\langle n_k \rangle^2$. We then find that the second-order correlation function may be expressed in terms of the first-order function as

$$G_{\mu_1\mu_2\mu_3\mu_4}{}^{(2)}(x_1 x_2, x_3 x_4) = G_{\mu_1\mu_3}{}^{(1)}(x_1,x_3) G_{\mu_2\mu_4}{}^{(1)}(x_2,x_4)$$

$$+ G_{\mu_1\mu_4}{}^{(1)}(x_1,x_4) G_{\mu_2\mu_3}{}^{(1)}(x_2,x_3). \quad (10.26)$$

It is easily shown that all of the higher order correlation functions as well reduce to sums of products of the first-order function. The nth-order correlation function may be written as

$$G_{\mu_1\cdots\mu_{2n}}{}^{(n)}(x_1\cdots x_n, x_{n+1}\cdots x_{2n}) = \sum_{\wp} \prod_{j=1}^{n} G_{\mu_j\nu_j}{}^{(1)}(x_j,y_j), \quad (10.27)$$

where the indices ν_j and the coordinates y_j for $j = 1 \cdots n$ are a permutation of the two sets $\mu_{n+1} \cdots \mu_{2n}$ and $x_{n+1} \cdots x_{2n}$, and the sum is carried out over all of the $n!$ permutations. One of the family resemblances which links all fields represented by the weight function (10.23) is that their properties may be fully described through knowledge of the first-order correlation function.

The fields which have traditionally been called coherent ones in optical terminology are easily described in terms of the first-order correlation function given by Eq. (10.25). Since the light in such fields is accurately collimated and nearly monochromatic, the mean occupation number $\langle n_{\mathbf{k},\lambda} \rangle$ vanishes outside a small volume of $\mathbf{k}$-space. The criterion for accurate coherence is ordinarily that the dimensions of this volume be extremely small in comparison to the magnitude of $\mathbf{k}$. It is easily verified, if the field is fully polarized, and the two points $(\mathbf{r},t)$ and $(\mathbf{r}',t')$ are not too distantly separated, that the correlation function (10.25) falls approximately into the factorized form of Eq. (2.4). That is to say, fields of the type we have described approximately fulfill the condition for first-order coherence.[3] It is easily seen, however, from the structure of the higher order correlation functions that these fields can never have second or higher order coherence. In fact, if we evaluate the function $G^{(n)}$ given by Eq. (10.27) for the particular case in which all of the coordinates are set equal, $x_1 = \cdots = x_{2n} = x$, and all of the indices as well, $\mu_1 = \cdots = \mu_{2n} = \mu$, we find the result

$$G_{\mu\cdots\mu}{}^{(n)}(x\cdots x, x\cdots x) = n! [G_{\mu\mu}{}^{(1)}(x,x)]^n. \quad (10.28)$$

The presence of the coefficient $n!$ in this expression is incompatible with the factorization condition (10.4) for the correlation functions of order n greater than one. The absence of second or higher order coherence is thus a general feature of stationary fields described by the Gaussian weight function (10.23). There exists, in other words, a fundamental sense in which these fields remain incoherent no matter how monochromatic or accurately collimated they are. We need hardly add that other types of fields such as those generated by radio transmitters or masers may possess arbitrarily high orders of coherence.

During the completion of the present paper a note by Sudarshan[12] has appeared which deals with some of the problems of photon statistics that have been treated here.[22] Sudarshan has observed the existence of what

---

[22] In an accompanying note, L. Mandel and E. Wolf [Phys. Rev. Letters **10**, 276 (1963)] warmly defend the classical approach to photon problems. Some of the possibilities and fundamental limitations of this approach should be evident from our earlier work. We may mention that the "implication" they draw from Ref. 1 and disagree with cannot be validly inferred from any reading of that paper.

we have called the P representation of the density operator and has stated its connection with the representation based on the n-quantum states. To that extent, his work agrees with ours in Secs. VII and IX. He has, however, made a number of statements which appear to attach an altogether different interpretation to the P representation. In particular, he regards its existence as demonstrating the "complete equivalence" of the classical and quantum mechanical approaches to photon statistics. He states further that there is a "one-to-one correspondence" between the weight functions P and the probability distributions for the field amplitudes of classical theory.

The relation between the P representation and classical theory has already been discussed at some length in Secs. VII–IX. We have shown there that the weight function $P(\alpha)$ is, in general, an intrinsically quantum-mechanical structure and not derivable from classical arguments. In the limit $\hbar \to 0$, which corresponds to large amplitudes of excitation for the modes, the weight functions $P(\alpha)$ may approach classical probability functions as asymptotic forms. Since infinitely many quantum states of the field may approach the same asymptotic form, it is clear that the correspondence between the weight functions $P(\alpha)$ and classical probability distributions is not at all one-to-one.

ACKNOWLEDGMENTS

The author is grateful to the Research Laboratory of the American Optical Company and its director, Dr. S. M. MacNeille, for partial support of this work.

논문 웹페이지

Controlling Electromagnetic Fields

J. B. Pendry,[1]* D. Schurig,[2] D. R. Smith[2]

Using the freedom of design that metamaterials provide, we show how electromagnetic fields can be redirected at will and propose a design strategy. The conserved fields—electric displacement field **D**, magnetic induction field **B**, and Poynting vector **B**—are all displaced in a consistent manner. A simple illustration is given of the cloaking of a proscribed volume of space to exclude completely all electromagnetic fields. Our work has relevance to exotic lens design and to the cloaking of objects from electromagnetic fields.

To exploit electromagnetism, we use materials to control and direct the fields: a glass lens in a camera to produce an image, a metal cage to screen sensitive equipment, "blackbodies" of various forms to prevent unwanted reflections. With homogeneous materials, optical design is largely a matter of choosing the interface between two materials. For example, the lens of a camera is optimized by altering its shape so as to minimize geometrical aberrations. Electromagnetically inhomogeneous materials offer a different approach to control light; the introduction of specific gradients in the refractive index of a material can be used to form lenses and other optical elements, although the types and ranges of such gradients tend to be limited.

[1]Department of Physics, Blackett Laboratory, Imperial College London, London SW7 2AZ, UK. [2]Department of Electrical and Computer Engineering, Duke University, Box 90291, Durham, NC 27708, USA.

*To whom correspondence should be addressed. E-mail: j.pendry@imperial.ac.uk

A new class of electromagnetic materials (1, 2) is currently under study: metamaterials, which owe their properties to subwavelength details of structure rather than to their chemical composition, can be designed to have properties difficult or impossible to find in nature. We show how the design flexibility of metamaterials can be used to achieve new electromagnetic devices and how metamaterials enable a new paradigm for the design of electromagnetic structures at all frequencies from optical down to DC.

Progress in the design of metamaterials has been impressive. A negative index of refraction (3) is an example of a material property that does not exist in nature but has been enabled by using metamaterial concepts. As a result, negative refraction has been much studied in recent years (4), and realizations have been reported at both GHz and optical frequencies (5–8). Novel magnetic properties have also been reported over a wide spectrum of frequencies. Further information on the design and construction of metamaterials may be found in (9–13). In fact, it is now conceivable that a material can be constructed whose permittivity and permeability values may be designed to vary independently and arbitrarily throughout a material, taking positive or negative values as desired.

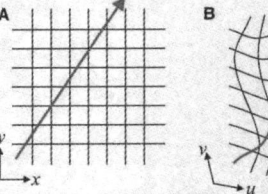

Fig. 1. (**A**) A field line in free space with the background Cartesian coordinate grid shown. (**B**) The distorted field line with the background coordinates distorted in the same fashion. The field in question may be the electric displacement or magnetic induction fields **D** or **B**, or the Poynting vector **S**, which is equivalent to a ray of light.

If we take this unprecedented control over the material properties and form inhomogeneous composites, we enable a powerful form of electromagnetic design. As an example of this design methodology, we show how the conserved quantities of electromagnetism—the electric displacement field **D**, the magnetic field intensity **B**, and the Poynting vector **S**—can all be directed at will, given access to the appropriate metamaterials. In particular, these fields can be focused as required or made to avoid objects and flow around them like a fluid, returning undisturbed to their original trajectories. These conclusions follow from exact manipulations of Maxwell's equations and are not confined to a ray approximation. They encompass in principle all forms of electromagnetic phenomena on all length scales.

We start with an arbitrary configuration of sources embedded in an arbitrary dielectric and magnetic medium. This initial configuration would be chosen to have the same topology as the final result we seek. For example, we might start with a uniform electric field and require that the field lines be moved to avoid a given region. Next, imagine the system is embedded in some elastic medium that can be pulled and stretched as we desire (Fig. 1). To keep track of distortions, we record the initial configuration of the fields on a Cartesian mesh, which is subsequently distorted by the same pulling and stretching process. The distortions can now be recorded as a coordinate transformation between the original Cartesian mesh and the distorted mesh

$$u(x,y,z), v(x,y,z), w(x,y,z) \quad (1)$$

where (u, v, w) is the location of the new point with respect to the x, y, and z axes. What happens to Maxwell's equations when we substitute the new coordinate system? The equations have exactly the same form in any coordinate system, but the refractive index—or more exactly the permittivity ε and permeability μ—are scaled by a common factor. In the new coordinate system,

we must use renormalized values of the permittivity and permeability:

$$\varepsilon'_u = \varepsilon_u \frac{Q_u Q_v Q_w}{Q_u^2},$$

$$\mu'_u = \mu_u \frac{Q_u Q_v Q_w}{Q_u^2}, \text{ etc.} \quad (2)$$

$$E'_u = Q_u E_u, H'_u = Q_u H_u, \text{ etc.} \quad (3)$$

where,

$$Q_u^2 = \left(\frac{\partial x}{\partial u}\right)^2 + \left(\frac{\partial y}{\partial u}\right)^2 + \left(\frac{\partial z}{\partial u}\right)^2$$

$$Q_v^2 = \left(\frac{\partial x}{\partial v}\right)^2 + \left(\frac{\partial y}{\partial v}\right)^2 + \left(\frac{\partial z}{\partial v}\right)^2$$

$$Q_w^2 = \left(\frac{\partial x}{\partial w}\right)^2 + \left(\frac{\partial y}{\partial w}\right)^2 + \left(\frac{\partial z}{\partial w}\right)^2 \quad (4)$$

As usual,

$$\mathbf{B}' = \mu_0 \mu' \mathbf{H}', \quad \mathbf{D}' = \varepsilon_0 \varepsilon' \mathbf{E}' \quad (5)$$

We have assumed orthogonal coordinate systems for which the formulae are particularly simple. The general case is given in (14) and in the accompanying online material (15). The equivalence of coordinate transformations and changes to ε and μ has also been referred to in (16).

Now let us put these transformations to use. Suppose we wish to conceal an arbitrary object contained in a given volume of space; furthermore, we require that external observers be unaware that something has been hidden from them. Our plan is to achieve concealment by cloaking the object with a metamaterial whose function is to deflect the rays that would have struck the object, guide them around the object, and return them to their original trajectory.

Our assumptions imply that no radiation can get into the concealed volume, nor can any radiation get out. Any radiation attempting to penetrate the secure volume is smoothly guided around by the cloak to emerge traveling in the same direction as if it had passed through the empty volume of space. An observer concludes that the secure volume is empty, but we are free to hide an object in the secure space. An alternative scheme has been recently investigated for the concealment of objects (17), but it relies on a specific knowledge of the shape and the material properties of the object being hidden. The electromagnetic cloak and the object concealed thus form a composite whose scattering properties can be reduced in the lowest order approximation: If the object changes, the cloak must change, too. In the scheme described here, an arbitrary object may be hidden because it remains untouched by external radiation. The method leads, in principle, to a perfect electromagnetic shield, excluding both propagating waves and near-fields from the concealed region.

For simplicity, we choose the hidden object to be a sphere of radius R_1 and the cloaking region to be contained within the annulus $R_1 < r < R_2$. A simple transformation that achieves the desired result can be found by taking all fields in the region $r < R_2$ and compressing them into the region $R_1 < r < R_2$,

$$r' = R_1 + r(R_2 - R_1)/R_2,$$

$$\theta' = \theta,$$

$$\phi' = \phi \quad (6)$$

Applying the transformation rules (15) gives the following values: for $r < R_1$, ε' and μ' are free to take any value without restriction and do not contribute to electromagnetic scattering; for $R_1 < r < R_2$

$$\varepsilon'_r = \mu'_r = \frac{R_2}{R_2 - R_1} \frac{(r' - R_1)^2}{r'},$$

$$\varepsilon'_\theta = \mu'_\theta = \frac{R_2}{R_2 - R_1},$$

$$\varepsilon'_\phi = \mu'_\phi = \frac{R_2}{R_2 - R_1} \quad (7)$$

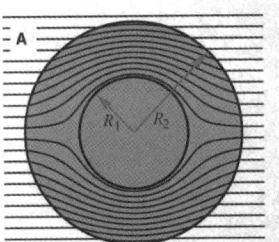
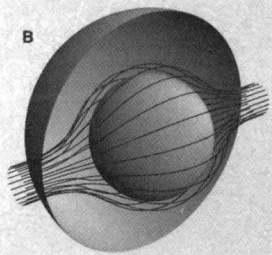
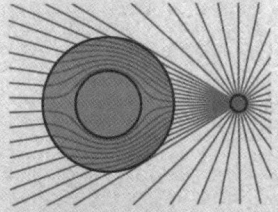

Fig. 2. A ray-tracing program has been used to calculate ray trajectories in the cloak, assuming that $R_2 \gg \lambda$. The rays essentially following the Poynting vector. **(A)** A two-dimensional (2D) cross section of rays striking our system, diverted within the annulus of cloaking material contained within $R_1 < r < R_2$ to emerge on the far side undeviated from their original course. **(B)** A 3D view of the same process.

Fig. 3. A point charge located near the cloaked sphere. We assume that $R_2 \ll \lambda$, the near-field limit, and plot the electric displacement field. The field is excluded from the cloaked region, but emerges from the cloaking sphere undisturbed. We plot field lines closer together near the sphere to emphasize the screening effect.

펜드리 논문 영문본

for $r > R_2$

$$\varepsilon'_{r'} = \mu'_{r'} = \varepsilon'_{\theta'} = \mu'_{\theta'} = \varepsilon'_{\phi'} = \mu'_{\phi'} = 1 \quad (8)$$

We stress that this prescription will exclude all fields from the central region. Conversely, no fields may escape from this region. At the outer surface of the cloak ($r = R_2$), we have $\varepsilon'_{r'} = \varepsilon'_{\theta'} = 1/\varepsilon'_{r'}$ and $\mu'_{\theta'} = \mu'_{\phi'} = 1/\mu'_{r'}$, which are the conditions for a perfectly matched layer (PML). Thus we can make the connection between this cloak, which is reflectionless by construction, and a well-studied reflectionless interface (*18*).

For purposes of illustration, suppose that $R_2 \gg \lambda$, where λ is the wavelength, so that we can use the ray approximation to plot the Poynting vector. If our system is then exposed to a source of radiation at infinity, we can perform the ray-tracing exercise shown in Fig. 2. Rays in this figure result from numerical integration of a set of Hamilton's equations obtained by taking the geometric limit of Maxwell's equations with anisotropic, inhomogeneous media. This integration provides independent confirmation that the configuration specified by Eqs. 6 and 7 excludes rays from the interior region. Alternatively, if $R_2 \ll \lambda$ and we locate a point charge nearby, the electrostatic (or magnetostatic) approximation applies. A plot of the local electrostatic displacement field is shown in Fig. 3.

Next we discuss the characteristics of the cloaking material. There is an unavoidable singularity in the ray tracing, as can be seen by considering a ray headed directly toward the center of the sphere (Fig. 2). This ray does not know whether to be deviated up or down, left or right. Neighboring rays are bent around in tighter and tighter arcs the closer to the critical ray they are. This in turn implies very rapid changes in ε' and μ', as sensed by the ray. These rapid changes are due (in a self-consistent way) to the tight turn of the ray and the anisotropy of ε' and μ'. Anisotropy of the medium is necessary because we have compressed space anisotropically.

Although anisotropy and even continuous variation of the parameters is not a problem for metamaterials (*19–21*), achieving very large or very small values of ε' and μ' can be. In practice, cloaking will be imperfect to the degree that we fail to satisfy Eq. 7. However, very considerable reductions in the cross section of the object can be achieved.

A further issue is whether the cloaking effect is broadband or specific to a single frequency. In the example we have given, the effect is only achieved at one frequency. This can easily be seen from the ray picture (Fig. 2). Each of the rays intersecting the large sphere is required to follow a curved, and therefore longer, trajectory than it would have done in free space, and yet we are requiring the ray to arrive on the far side of the sphere with the same phase. This implies a phase velocity greater that the velocity of light in vacuum which violates no physical law.

However, if we also require absence of dispersion, the group and phase velocities will be identical, and the group velocity can never exceed the velocity of light. Hence, in this instance the cloaking parameters must disperse with frequency and therefore can only be fully effective at a single frequency. We mention in passing that the group velocity may sometimes exceed the velocity of light (*22*) but only in the presence of strong dispersion. On the other hand, if the system is embedded in a medium having a large refractive index, dispersion may in principle be avoided and the cloaking operate over a broad bandwidth.

We have shown how electromagnetic fields can be dragged into almost any desired configuration. The distortion of the fields is represented as a coordinate transformation, which is then used to generate values of electrical permittivity and magnetic permeability ensuring that Maxwell's equations are still satisfied. The new concept of metamaterials is invoked, making realization of these designs a practical possibility.

References and Notes

1. J. B. Pendry, A. J. Holden, W. J. Stewart, I. Youngs, *Phys. Rev. Lett.* **76**, 4773 (1996).
2. J. B. Pendry, A. J. Holden, D. J. Robbins, W. J. Stewart, *IEEE Trans. Micr. Theory Techniques* **47**, 2075 (1999).
3. V. G. Veselago, *Soviet Physics USPEKI* **10**, 509 (1968).
4. D. R. Smith, W. J. Padilla, D. C. Vier, S. C. Nemat-Nasser, S. Schultz, *Phys. Rev. Lett.* **84**, 4184 (2000).
5. R. A. Shelby, D. R. Smith, S. Schultz, *Science* **292**, 77 (2001).
6. A. A. Houck, J. B. Brock, I. L. Chuang, *Phys. Rev. Lett.* **90**, 137401 (2003).
7. A. Grbic, G. V. Eleftheriades, *Phys. Rev. Lett.* **92**, 117403 (2004).
8. V. M. Shalaev et al., *Opt. Lett.* **30**, 3356 (2005).
9. D. R. Smith, J. B. Pendry, M. C. K. Wiltshire, *Science* **305**, 788 (2004).
10. E. Cubukcu, K. Aydin, E. Ozbay, S. Foteinopoulou, C. M. Soukoulis, *Nature* **423**, 604 (2003).
11. E. Cubukcu, K. Aydin, E. Ozbay, S. Foteinopoulou, C. M. Soukoulis, *Phys. Rev. Lett.* **91**, 207401 (2003).
12. T. J. Yen et al., *Science* **303**, 1494 (2004).
13. S. Linden et al., *Science* **306**, 1351 (2004).
14. A. J. Ward, J. B. Pendry, *J. Mod. Opt.* **43**, 773 (1996).
15. Methods are available as supporting material on *Science* Online.
16. U. Leonhardt, *IEEE J. Selected Topics Quantum Electronics* **9**, 102 (2003).
17. A. Alu, N. Engheta, *Phys. Rev.* **E95**, 016623 (2005).
18. J.-P. Berenger, *J. Comput. Phys.* **114**, 185 (1994).
19. D. R. Smith, J. J. Mock, A. F. Starr, D. Schurig, *Phys. Rev. E* **71**, 036617 (2005).
20. T. Driscoll et al., *Appl. Phys. Lett.* **88**, 081101 (2006).
21. R. B. Greegor et al., *Appl. Phys. Lett.* **87**, 091114 (2005).
22. R. Y. Chiao, P. W. Milonni, *Optics and Photonics News*, June (2002).
23. J.B.P. thanks the Engineering and Physical Sciences Research Council (EPSRC) for a Senior Fellowship, the European Community (EC) under project FP6-NMP4-CT-2003-505699, Department of Defense Office of Naval Research (DOD/ONR) Multidisciplinary Research Program of the University Research Institute (MURI) grant N00014-01-1-0803, DOD/ONR grant N00014-05-1-0861, and the EC Information Societies Technology (IST) program Development and Analysis of Left-Handed Materials (DALHM), project number IST-2001-35511, for financial support. D. Schurig acknowledges support from the Intelligence Community (IC) Postdoctoral Fellowship Program.

Supporting Online Material
www.sciencemag.org/cgi/content/full/1125907/DC1
SOM Text
Figs. S1 to S3

7 February 2006; accepted 26 April 2006
Published online 25 May 2006;
10.1126/science.1125907
Include this information when citing this paper.

논문 웹페이지

위대한 논문과의 만남을 마무리하며

    이 책은 양자광학의 창시자인 글라우버의 1963년 논문과 이 논문이 나오는 데 큰 역할을 했던 다른 과학자들의 논문에 초점을 맞추었습니다.

    양자광학을 공부하기 앞서 고전광학에서 빛의 파동성과 입자성에 관한 역사를 재고찰했고, 광통신의 역사와 광섬유의 발명에 얽힌 이야기로 책을 시작했습니다. 플랑크의 광자 이론과 아인슈타인의 레이저 이론 소개와 함께 메이저와 레이저의 발명에 관한 재미있는 이야기도 다루었습니다.

    이어서 디랙이 소개한 광자의 생성 및 소멸 연산자 아이디어를 전자기파에 대응한 이론에 대해 설명했습니다. 이 과정을 이해하기 위해서는 맥스웰 방정식을 먼저 알아야 합니다. 맥스웰 방정식의 자세한 내용은 이 시리즈의 《특수상대성이론》을 참고하기 바랍니다.

    다음으로는 빛과 물질의 상호작용에 관한 제인스와 커밍스의 연구 결과를 독자들이 이해할 수 있도록 간단한 수식만을 사용해 다루었습니다. 그리고 글라우버가 도입한 결맞는 상태와 반응집광의 개념을 이야기했습니다.

마지막으로 양자광학을 이용한 양자광 기술을 소개했습니다. 광학 핀셋, 주파수 빗, 메타물질을 이용한 투명 망토, 레이저 냉각, 양자정보과학에의 활용 등을 다루었습니다. 이 책을 통해 독자들이 양자광학의 신비에 푹 빠질 수 있으리라 생각합니다.

이 책의 출판 기획상 수식을 피할 수 없을 때는 고등학교 수학 정도를 아는 사람이라면 이해할 수 있도록 처음 쓴 원고를 고치고 또 고치는 작업을 반복했습니다. 그렇게 하여 수식을 줄여보려고 했습니다. 하지만 물리를 좋아하는 사람들이 쉽게 따라갈 수 있도록 친절하게 설명했습니다.

원고를 쓰기 위해 20세기의 여러 논문을 뒤적거렸습니다. 지금과는 완연히 다른 용어와 기호 때문에 많이 힘들었습니다. 특히 번역이 안 되어 있는 자료들이 많았지만 프랑스 논문에 대해서는 불문과를 졸업한 아내의 도움으로 조금은 이해할 수 있었습니다.

집필을 끝내자마자 다시 양자물질에 대한 오리지널 논문을 공부하며, 시리즈를 계속 이어나갈 생각을 하니 즐거움에 벅차오릅니다. 제가 느끼는 이 기쁨을 독자들이 공유할 수 있기를 바라며 이제 힘들었지만 재미있었던 양자광학 논문과의 씨름을 여기서 멈추려고 합니다.

끝으로 용기를 내서 이 시리즈의 출간을 결정한 성림원북스의 이

성림 사장과 직원들에게 감사를 드립니다. 시리즈 초안이 나왔을 때, 수식이 많아 출판사들이 꺼릴 것 같다는 생각이 들었습니다. 몇 군데에 출판을 의뢰한 후 거절당하면 블로그에 올릴 생각으로 글을 써 내려갔습니다. 놀랍게도 첫 번째로 이 원고의 이야기를 나눈 성림원북스에서 출간을 결정해 주어서 책이 나올 수 있게 되었습니다. 원고를 쓰는 데 필요한 프랑스 논문의 번역을 도와준 아내에게도 고마움을 전합니다. 그리고 이 책을 쓸 수 있도록 멋진 논문을 만든 글라우버 박사님에게도 감사를 드립니다.

진주에서 정완상 교수

이 책을 위해 참고한 논문들

1장

[1] I. Newton, Opticks: or, a treatise of the reflexions, refractions, inflexions and colours of light, 1704.

[2] C. Huygens, Traité de la Lumière, 1678.

2장

[1] M. Planck, "Über eine Verbesserung der Wienschen Spektralgleichung", Verhandlungen der Deutschen Physikalischen Gesellschaft. 2; 202, 1900.

[2] M. Planck, "Zur Theorie des Gesetzes der Energieverteilung im Normalspectrum", Verhandlungen der Deutschen Physikalischen Gesellschaft. 2; 237, 1900.

[3] M. Planck, "Entropie und Temperatur strahlender Wärme", Annalen der Physik. 306; 719, 1900.

[4] A. Einstein, "Über einen die Erzeugung und Verwandlung des Lichtes betreffenden heuristischen Gesichtspunkt", Annalen der Physik (in German). 17 (6): 132-148, 1905.

[5] Arthur H. Compton, "A Quantum Theory of the Scattering of X-Rays by Light Elements", Physical Review. 21 (5): 483-

502, May 1923.

[6] N. Bohr, "On the Constitution of Atoms and Molecules", Philosophical Magazinc. 26 (151): 1-24, 1913.

[7] A. Einstein, "Zur Quantentheorie der Strahlung", Physikalische Zeitschrift. 18: 121-128, 1917.

[8] W. Heisenberg, "Über quantentheoretische Umdeutung kinematischer und mechanischer Beziehungen", Zeitschrift für Physik. 33 (1): 879-893, 1925.

[9] E. Schrödinger, An Undulatory Theory of the Mechanics of Atoms and Molecules, Phys. Rev. 28; 1049, 1926.

[10] M. Born and P. Jordan, "Zur Quantenmechanik", Zeitschrift für Physik. 34 (1): 858-888, 1925.

[11] P. A. M. Dirac, "On the Theory of Quantum Mechanics", Proceedings of the Royal Society A. 112 (762): 661-677, 1926.

[12] A. L. Schawlow and C. H. Townes, "Infrared and optical masers", Physical Review. 112 (6-15): 1940-1949.

[13] T. H. Maiman, "Stimulated Optical Radiation in Ruby", Nature. 187 (4736): 493-494, 1960.

3장

[1] H. H. Hopkins and N. S. Kapany, "A flexible fibrescope, using static scanning", Nature. 173 (4392): 39-41, 1954.

[2] K. C. Kao and G. A. Hockham, "Dielectric-fibre surface waveguides for optical frequencies," Proc. IEE, Vol. 113, pp. 1151-1158, 1966.

4장

[1] P. A. M. Dirac, "The Quantum Theory of the Emission and Absorption of Radiation", Proceedings of the Royal Society A. 114 (767): 243-265, 1927.

5장

[1] E. T. Jaynes and F. W. Cummings, "Comparison of quantum and semiclassical radiation theories with application to the beam maser", Proc. IEEE. 51 (1): 89-109, 1963.

6장

[1] R. J. Glauber, "Coherent and Incoherent States of the Radiation Field", Physical Review. 131 (6). American Physical Society (APS): 2766-2788.

7장

[1] A. Ashkin, "Acceleration and Trapping of Particles by Radiation Pressure", Physical Review Letters. 24 (4): 156-159,

1970.

[2] A. Ashkin, J. M. Dziedzic, J. E. Bjorkholm and S. Chu, "Observation of a single-beam gradient force optical trap for dielectric particles", Optics Letters. 11 (5): 288-290, 1986.

[3] V. G. Veselago, "The electrodynamics of substances with simultaneously negative values of ε and μ", Sov. Phys. Usp. 10 (4): 509-514, 1967.

[4] J. B. Pendry, "Negative Refraction Makes a Perfect Lens", Physical Review Letters. 85 (18): 3966-3969, 2000.

[5] J. B. Pendry, D. Schurig and D. R. Smith, "Controlling Electromagnetic Fields", Science. 312 (5781): 1780-1782, 2006.

수식에 사용하는 그리스 문자

대문자	소문자	읽기	대문자	소문자	읽기
A	α	알파(alpha)	N	ν	뉴(nu)
B	β	베타(beta)	Ξ	ξ	크시(xi)
Γ	γ	감마(gamma)	O	o	오미크론(omicron)
Δ	δ	델타(delta)	Π	π	파이(pi)
E	ε	엡실론(epsilon)	P	ρ	로(rho)
Z	ζ	제타(zeta)	Σ	σ	시그마(sigma)
H	η	에타(eta)	T	τ	타우(tau)
Θ	θ	세타(theta)	Y	υ	입실론(upsilon)
I	ι	요타(iota)	Φ	φ	피(phi)
K	$\varkappa$	카파(kappa)	X	χ	키(chi)
Λ	λ	람다(lambda)	Ψ	ψ	프시(psi)
M	μ	뮤(mu)	Ω	ω	오메가(omega)

노벨 물리학상 수상자들을 소개합니다

이 책에 언급된 노벨상 수상자는 이름 앞에 ★로 표시하였습니다.

연도	수상자	수상 이유
1901	빌헬름 콘라트 뢴트겐	그의 이름을 딴 놀라운 광선의 발견으로 그가 제공한 특별한 공헌을 인정하여
1902	헨드릭 안톤 로런츠 피터르 제이만	복사 현상에 대한 자기의 영향에 대한 연구를 통해 그들이 제공한 탁월한 공헌을 인정하여
1903	앙투안 앙리 베크렐	자발 방사능 발견으로 그가 제공한 탁월한 공로를 인정하여
	피에르 퀴리 마리 퀴리	앙리 베크렐 교수가 발견한 방사선 현상에 대한 공동 연구를 통해 그들이 제공한 탁월한 공헌을 인정하여
1904	존 윌리엄 스트럿 레일리	가장 중요한 기체의 밀도에 대한 조사와 이러한 연구와 관련하여 아르곤을 발견한 공로
1905	필리프 레나르트	음극선에 대한 연구
1906	조지프 존 톰슨	기체에 의한 전기 전도에 대한 이론적이고 실험적인 연구의 큰 장점을 인정하여
1907	앨버트 에이브러햄 마이컬슨	광학 정밀 기기와 그 도움으로 수행된 분광 및 도량형 조사
1908	가브리엘 리프만	간섭 현상을 기반으로 사진적으로 색상을 재현하는 방법
1909	굴리엘모 마르코니 카를 페르디난트 브라운	무선 전신 발전에 기여한 공로를 인정받아
1910	유하네스 디데릭 판데르발스	기체와 액체의 상태 방정식에 관한 연구
1911	빌헬름 빈	열복사 법칙에 관한 발견
1912	닐스 구스타프 달렌	등대와 부표를 밝히기 위해 가스 어큐뮬레이터와 함께 사용하기 위한 자동 조절기 발명

노벨 물리학상 수상자 목록

1913	헤이커 카메를링 오너스	특히 액체 헬륨 생산으로 이어진 저온에서의 물질 특성에 대한 연구
1914	막스 폰 라우에	결정에 의한 X선 회절 발견
1915	윌리엄 헨리 브래그 윌리엄 로런스 브래그	X선을 이용한 결정구조 분석에 기여한 공로
1916	수상자 없음	
1917	찰스 글러버 바클라	원소의 특징적인 뢴트겐 복사 발견
1918	★막스 플랑크	에너지 양자 발견으로 물리학 발전에 기여한 공로 인정
1919	요하네스 슈타르크	커낼선의 도플러 효과와 전기장에서 분광선의 분할 발견
1920	샤를 에두아르 기욤	니켈강 합금의 이상 현상을 발견하여 물리학의 정밀 측정에 기여한 공로를 인정하여
1921	★알베르트 아인슈타인	이론 물리학에 대한 공로, 특히 광전효과 법칙 발견
1922	★닐스 보어	원자 구조와 원자에서 방출되는 방사선 연구에 기여
1923	로버트 앤드루스 밀리컨	전기의 기본 전하와 광전효과에 관한 연구
1924	칼 만네 예오리 시그반	X선 분광학 분야에서의 발견과 연구
1925	제임스 프랑크 구스타프 헤르츠	전자가 원자에 미치는 영향을 지배하는 법칙 발견
1926	장 바티스트 페랭	물질의 불연속 구조에 관한 연구, 특히 침전 평형 발견
1927	★아서 콤프턴	그의 이름을 딴 효과 발견
	찰스 톰슨 리스 윌슨	수증기 응축을 통해 전하를 띤 입자의 경로를 볼 수 있게 만든 방법
1928	오언 윌런스 리처드슨	열전자 현상에 관한 연구, 특히 그의 이름을 딴 법칙 발견
1929	루이 드브로이	전자의 파동성 발견
1930	찬드라세카라 벵카타 라만	빛의 산란에 관한 연구와 그의 이름을 딴 효과 발견
1931	수상자 없음	

1932	★베르너 하이젠베르크	수소의 동소체 형태 발견으로 이어진 양자역학의 창시
1933	★에르빈 슈뢰딩거 ★폴 디랙	원자 이론의 새로운 생산적 형태 발견
1934	수상자 없음	
1935	제임스 채드윅	중성자 발견
1936	빅토르 프란츠 헤스	우주 방사선 발견
	칼 데이비드 앤더슨	양전자 발견
1937	클린턴 조지프 데이비슨 조지 패짓 톰슨	결정에 의한 전자의 회절에 대한 실험적 발견
1938	엔리코 페르미	중성자 조사에 의해 생성된 새로운 방사성 원소의 존재에 대한 시연 및 이와 관련된 느린중성자에 의한 핵반응 발견
1939	어니스트 로런스	사이클로트론의 발명과 개발, 특히 인공 방사성 원소와 관련하여 얻은 결과
1940	수상자 없음	
1941		
1942		
1943	오토 슈테른	분자선 방법 개발 및 양성자의 자기 모멘트 발견에 기여
1944	★이지도어 아이작 라비	원자핵의 자기적 특성을 기록하기 위한 공명 방법
1945	볼프강 파울리	파울리 원리라고도 불리는 배제 원리의 발견
1946	퍼시 윌리엄스 브리지먼	초고압을 발생시키는 장치의 발명과 고압 물리학 분야에서 그가 이룬 발견에 대해
1947	에드워드 빅터 애플턴	대기권 상층부의 물리학 연구, 특히 이른바 애플턴층의 발견
1948	패트릭 메이너드 스튜어트 블래킷	윌슨 구름상자 방법의 개발과 핵물리학 및 우주 방사선 분야에서의 발견
1949	유카와 히데키	핵력에 관한 이론적 연구를 바탕으로 중간자 존재 예측

1950	세실 프랭크 파월	핵 과정을 연구하는 사진 방법의 개발과 이 방법으로 만들어진 중간자에 관한 발견
1951	존 더글러스 콕크로프트	인위적으로 가속된 원자 입자에 의한 원자핵 변환에 대한 선구자적 연구
	어니스트 토머스 신턴 월턴	
1952	펠릭스 블로흐	핵자기 정밀 측정을 위한 새로운 방법 개발 및 이와 관련된 발견
	에드워드 밀스 퍼셀	
1953	프리츠 제르니커	위상차 방법 시연, 특히 위상차 현미경 발명
1954	★막스 보른	양자역학의 기초 연구, 특히 파동함수의 통계적 해석
	발터 보테	우연의 일치 방법과 그 방법으로 이루어진 그의 발견
1955	★윌리스 유진 램	수소 스펙트럼의 미세 구조에 관한 발견
	폴리카프 쿠시	전자의 자기 모멘트를 정밀하게 측정한 공로
1956	윌리엄 브래드퍼드 쇼클리	반도체 연구 및 트랜지스터 효과 발견
	존 바딘	
	월터 하우저 브래튼	
1957	양전닝	소립자에 관한 중요한 발견으로 이어진 소위 패리티 법칙에 대한 철저한 조사
	리정다오	
1958	파벨 알렉세예비치 체렌코프	체렌코프 효과의 발견과 해석
	일리야 프란크	
	이고리 탐	
1959	에밀리오 지노 세그레	반양성자 발견
	오언 체임벌린	
1960	도널드 아서 글레이저	거품 상자의 발명
1961	로버트 호프스태터	원자핵의 전자 산란에 대한 선구적인 연구와 핵자 구조에 관한 발견
	루돌프 뫼스바워	감마선의 공명 흡수에 관한 연구와 그의 이름을 딴 효과에 대한 발견

1962	레프 다비도비치 란다우	응집 물질, 특히 액체 헬륨에 대한 선구적인 이론
1963	유진 폴 위그너	원자핵 및 소립자 이론에 대한 공헌, 특히 기본 대칭 원리의 발견 및 적용을 통한 공로
	마리아 괴페르트 메이어	핵 껍질 구조에 관한 발견
	한스 옌센	
1964	★니콜라이 바소프	메이저-레이저 원리에 기반한 발진기 및 증폭기의 구성으로 이어진 양자 전자 분야의 기초 작업
	★알렉산드르 프로호로프	
	★찰스 하드 타운스	
1965	도모나가 신이치로	소립자의 물리학에 심층적인 결과를 가져온 양자전기역학의 근본적인 연구
	★줄리언 슈윙거	
	리처드 필립스 파인먼	
1966	★알프레드 카스틀레르	원자에서 헤르츠 공명을 연구하기 위한 광학적 방법의 발견 및 개발
1967	한스 알브레히트 베테	핵반응 이론, 특히 별의 에너지 생산에 관한 발견에 기여
1968	루이스 월터 앨버레즈	소립자 물리학에 대한 결정적인 공헌, 특히 수소 기포 챔버 사용 기술 개발과 데이터 분석을 통해 가능해진 다수의 공명 상태 발견
1969	머리 겔만	기본 입자의 분류와 그 상호 작용에 관한 공헌 및 발견
1970	한네스 올로프 예스타 알벤	플라스마 물리학의 다양한 부분에서 유익한 응용을 통해 자기유체역학의 기초 연구 및 발견
	루이 외젠 펠릭스 네엘	고체물리학에서 중요한 응용을 이끈 반강자성 및 강자성에 관한 기초 연구 및 발견
1971	데니스 가보르	홀로그램 방법의 발명 및 개발
1972	존 바딘	일반적으로 BCS 이론이라고 하는 초전도 이론을 공동으로 개발한 공로
	리언 닐 쿠퍼	
	존 로버트 슈리퍼	

연도	수상자	업적
1973	에사키 레오나	반도체와 초전도체의 터널링 현상에 관한 실험적 발견
	이바르 예베르	
	브라이언 데이비드 조지프슨	터널 장벽을 통과하는 초전류 특성, 특히 일반적으로 조지프슨 효과로 알려진 현상에 대한 이론적 예측
1974	마틴 라일	전파 천체물리학의 선구적인 연구: 라일은 특히 개구 합성 기술의 관찰과 발명, 그리고 휴이시는 펄서 발견에 결정적인 역할을 함
	앤터니 휴이시	
1975	오게 닐스 보어	원자핵에서 집단 운동과 입자 운동 사이의 연관성 발견과 이 연관성에 기초한 원자핵 구조 이론 개발
	벤 로위 모텔손	
	제임스 레인워터	
1976	버턴 릭터	새로운 종류의 무거운 기본 입자 발견에 대한 선구적인 작업
	새뮤얼 차오 충 팅	
1977	필립 워런 앤더슨	자기 및 무질서 시스템의 전자 구조에 대한 근본적인 이론적 조사
	네빌 프랜시스 모트	
	존 해즈브룩 밴블렉	
1978	표트르 레오니도비치 카피차	저온 물리학 분야의 기본 발명 및 발견
	아노 앨런 펜지어스	우주 마이크로파 배경 복사의 발견
	로버트 우드로 윌슨	
1979	셸던 리 글래쇼	특히 약한 중성 전류의 예측을 포함하여 기본 입자 사이의 통일된 약한 전자기 상호 작용 이론에 대한 공헌
	압두스 살람	
	스티븐 와인버그	
1980	제임스 왓슨 크로닌	중성 K 중간자의 붕괴에서 기본 대칭 원리 위반 발견
	밸 로그즈던 피치	
1981	니콜라스 블룸베르헌	레이저 분광기 개발에 기여
	★아서 레너드 숄로	
	카이 만네 뵈리에 시그반	고해상도 전자 분광기 개발에 기여

연도	수상자	업적
1982	케네스 게디스 윌슨	상전이와 관련된 임계 현상에 대한 이론
1983	수브라마니안 찬드라세카르	별의 구조와 진화에 중요한 물리적 과정에 대한 이론적 연구
	윌리엄 앨프리드 파울러	우주의 화학 원소 형성에 중요한 핵반응에 대한 이론 및 실험적 연구
1984	카를로 루비아	약한 상호 작용의 커뮤니케이터인 필드 입자 W와 Z의 발견으로 이어진 대규모 프로젝트에 결정적인 기여
	시몬 판데르 메이르	
1985	클라우스 폰 클리칭	양자화된 홀 효과의 발견
1986	에른스트 루스카	전자 광학의 기초 작업과 최초의 전자 현미경 설계
	게르트 비니히	스캐닝 터널링 현미경 설계
	하인리히 로러	
1987	요하네스 게오르크 베드노르츠	세라믹 재료의 초전도성 발견에서 중요한 돌파구
	카를 알렉산더 뮐러	
1988	리언 레더먼	뉴트리노 빔 방법과 뮤온 중성미자 발견을 통한 경입자의 이중 구조 증명
	멜빈 슈워츠	
	잭 스타인버거	
1989	노먼 포스터 램지	분리된 진동 필드 방법의 발명과 수소 메이저 및 기타 원자시계에서의 사용
	한스 게오르크 데멜트	이온 트랩 기술 개발
	볼프강 파울	
1990	제롬 프리드먼	입자 물리학에서 쿼크 모델 개발에 매우 중요한 역할을 한 양성자 및 구속된 중성자에 대한 전자의 심층 비탄성 산란에 관한 선구적인 연구
	헨리 웨이 켄들	
	리처드 테일러	
1991	피에르질 드 젠	간단한 시스템에서 질서 현상을 연구하기 위해 개발된 방법을 보다 복잡한 형태의 물질, 특히 액정과 고분자로 일반화할 수 있음을 발견

연도	수상자	업적
1992	조르주 샤르파크	입자 탐지기, 특히 다중 와이어 비례 챔버의 발명 및 개발
1993	러셀 헐스	새로운 유형의 펄서 발견, 중력 연구의 새로운 가능성을 연 발견
	조지프 테일러	
1994	버트럼 브록하우스	중성자 분광기 개발
	클리퍼드 셜	중성자 회절 기술 개발
1995	마틴 펄	타우 렙톤의 발견
	프레더릭 라이너스	중성미자 검출
1996	데이비드 리	헬륨-3의 초유동성 발견
	더글러스 오셔로프	
	로버트 리처드슨	
1997	★스티븐 추	레이저 광으로 원자를 냉각하고 가두는 방법 개발
	★클로드 코엔타누지	
	★윌리엄 필립스	
1998	로버트 로플린	부분적으로 전하를 띤 새로운 형태의 양자 유체 발견
	호르스트 슈퇴르머	
	대니얼 추이	
1999	헤라르뒤스 엇호프트	물리학에서 전기약력 상호작용의 양자 구조 규명
	마르티뉘스 펠트만	
2000	조레스 알표로프	정보 통신 기술에 대한 기초 작업(고속 및 광전자 공학에 사용되는 반도체 이종 구조 개발)
	허버트 크로머	
	잭 킬비	정보 통신 기술에 대한 기초 작업(집적회로 발명에 기여)
2001	에릭 코넬	알칼리 원자의 희석 가스에서 보스-아인슈타인 응축 달성 및 응축 특성에 대한 초기 기초 연구
	칼 위먼	
	볼프강 케테를레	

연도	수상자	업적
2002	레이먼드 데이비스	천체물리학, 특히 우주 중성미자 검출에 대한 선구적인 공헌
	고시바 마사토시	
	리카르도 자코니	우주 X선 소스의 발견으로 이어진 천체물리학에 대한 선구적인 공헌
2003	알렉세이 아브리코소프	초전도체 및 초유체 이론에 대한 선구적인 공헌
	비탈리 긴즈부르크	
	앤서니 레깃	
2004	데이비드 그로스	강한 상호작용 이론에서 점근적 자유의 발견
	데이비드 폴리처	
	프랭크 윌첵	
2005	★로이 글라우버	광학 일관성의 양자 이론에 기여
	★존 홀	광 주파수 콤 기술을 포함한 레이저 기반 정밀 분광기 개발에 기여
	★테오도어 헨슈	
2006	존 매더	우주 마이크로파 배경 복사의 흑체 형태와 이방성 발견
	조지 스무트	
2007	알베르 페르	자이언트 자기 저항의 발견
	페터 그륀베르크	
2008	난부 요이치로	아원자 물리학에서 자발적인 대칭 깨짐 메커니즘 발견
	고바야시 마코토	자연계에 적어도 세 종류의 쿼크가 존재함을 예측하는 깨진 대칭의 기원 발견
	마스카와 도시히데	
2009	★찰스 가오	광 통신을 위한 섬유의 빛 전송에 관한 획기적인 업적
	윌러드 보일	영상 반도체 회로(CCD 센서)의 발명
	조지 엘우드 스미스	
2010	안드레 가임	2차원 물질 그래핀에 관한 획기적인 실험
	콘스탄틴 노보셀로프	

연도	수상자	업적
2011	솔 펄머터 브라이언 슈밋 애덤 리스	원거리 초신성 관측을 통한 우주 가속 팽창 발견
2012	세르주 아로슈 데이비드 와인랜드	개별 양자 시스템의 측정 및 조작을 가능하게 하는 획기적인 실험 방법
2013	프랑수아 앙글레르 피터 힉스	아원자 입자의 질량 기원에 대한 이해에 기여하고 최근 CERN의 대형 하드론 충돌기에서 ATLAS 및 CMS 실험을 통해 예측된 기본 입자의 발견을 통해 확인된 메커니즘의 이론적 발견
2014	아카사키 이사무 아마노 히로시 나카무라 슈지	밝고 에너지 절약형 백색 광원을 가능하게 한 효율적인 청색 발광 다이오드의 발명
2015	가지타 다카아키 아서 맥도널드	중성미자가 질량을 가지고 있음을 보여주는 중성미자 진동 발견
2016	데이비드 사울레스 덩컨 홀데인 마이클 코스털리츠	위상학적 상전이와 물질의 위상학적 위상에 대한 이론적 발견
2017	라이너 바이스 킵 손 배리 배리시	LIGO 탐지기와 중력파 관찰에 결정적인 기여
2018	★아서 애슈킨	레이저 물리학 분야의 획기적인 발명(광학 핀셋과 생물학적 시스템에 대한 응용)
	제라르 무루 도나 스트리클런드	레이저 물리학 분야의 획기적인 발명(고강도 초단파 광 펄스 생성 방법)
2019	제임스 피블스	우주의 진화와 우주에서 지구의 위치에 대한 이해에 기여(물리 우주론의 이론적 발견)
	미셸 마요르 디디에 쿠엘로	우주의 진화와 우주에서 지구의 위치에 대한 이해에 기여(태양형 항성 주위를 공전하는 외계 행성 발견)

연도	수상자	업적
2020	로저 펜로즈	블랙홀 형성이 일반 상대성 이론의 확고한 예측이라는 발견
	라인하르트 겐첼	우리 은하의 중심에 있는 초거대 밀도 물체 발견
	앤드리아 게즈	
2021	마나베 슈쿠로	복잡한 시스템에 대한 이해에 획기적인 기여(지구 기후의 물리적 모델링, 가변성을 정량화하고 지구 온난화를 안정적으로 예측)
	클라우스 하셀만	
	조르조 파리시	복잡한 시스템에 대한 이해에 획기적인 기여 (원자에서 행성 규모에 이르는 물리적 시스템의 무질서와 요동의 상호작용 발견)
2022	★알랭 아스페	얽힌 광자를 사용한 실험, 벨 불평등 위반 규명 및 양자 정보 과학 개척
	존 클라우저	
	안톤 차일링거	
2023	피에르 아고스티니	물질의 전자 역학 연구를 위해 아토초(100경분의 1초) 빛 펄스를 생성하는 실험 방법 고안
	페렌츠 크러우스	
	안 륄리에	
2024	존 홉필드	인공신경망을 이용해 머신러닝을 가능하게 하는 기초적인 발견과 발명
	제프리 힌턴